Demon
of the Waters

This Large Print Book carries the
Seal of Approval of N.A.V.H.

Demon

of the Waters

*The True Story of the Mutiny
on the Whaleship Globe*

Gregory Gibson

Illustrations by
Erik Ronnberg *and* **Gary Tonkin**

Thorndike Press • Waterville, Maine

Published in 2002 by arrangement with Little, Brown and Company, Inc.

Thorndike Press Large Print Adventure Series.

The tree indicium is a trademark of Thorndike Press.

The text of this Large Print edition is unabridged.
Other aspects of the book may vary from the original edition.

Cover design by William B. Hubschwerlin.

Set in 16 pt. Plantin.

Printed in the United States on permanent paper.

Library of Congress Cataloging-in-Publication Data

Gibson, Gregory, 1945–
    Demon of the waters : the true story of the mutiny on the whaleship Globe / Gregory Gibson ; illustrations by Erik Ronnberg and Gary Tonkin.
        p. cm.
    Originally published: 1st ed. Boston : Little, Brown, c2002.
    Includes bibliographical references (p. 514).
    ISBN 0-7862-4459-3 (lg. print : hc : alk. paper)
    1. Globe (Whaling ship) 2. Globe Mutiny, 1824. I. Title.
G545 .G53 2002
910.4´5—dc21   2002020389

For Celia, Broox, and Anne Marie

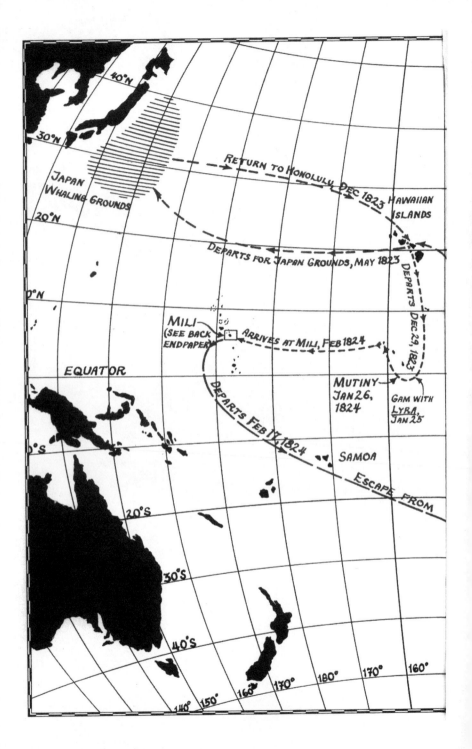

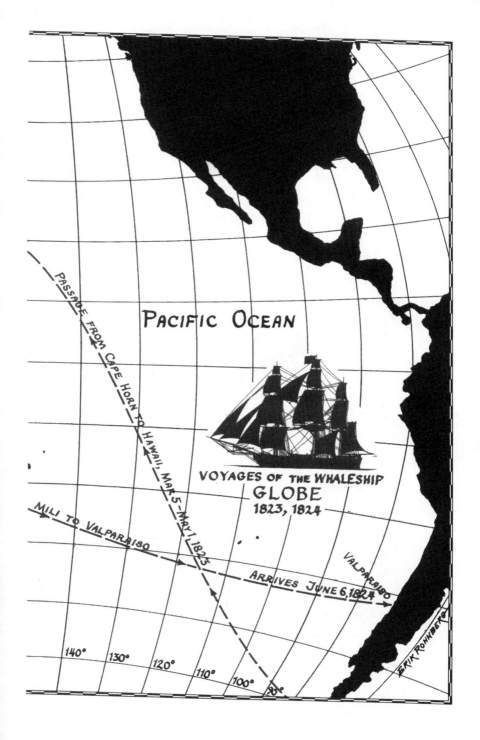

PACIFIC OCEAN

PASSAGE FROM CAPE HORN TO HAWAII, MAR.5—MAY 1,1823

MILI TO VALPARAISO

VOYAGES OF THE WHALESHIP
GLOBE
1823, 1824

VALPARAISO

ARRIVES JUNE 6,1824

140° 130° 120° 110° 100° 90°

ERIK RONNBERG

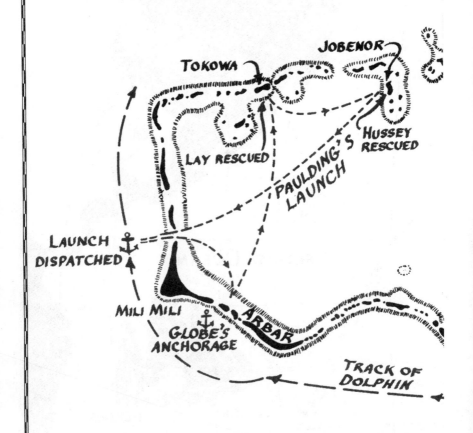

JOBENOR

TOKOWA

HUSSEY
RESCUED

LAY RESCUED

PAULDING'S
LAUNCH

LAUNCH
DISPATCHED

MILI MILI

ARBAR

GLOBE'S
ANCHORAGE

TRACK OF
DOLPHIN

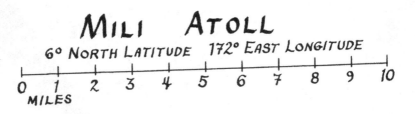

MILI ATOLL
6° NORTH LATITUDE   172° EAST LONGITUDE

0   1   2   3   4   5   6   7   8   9   10
MILES

HOMER'S MARCH
BEGINS

LUKUNOR

ATTEMPT AT
LUKUNOR PASSAGE

CHIRUBON

FIRST LANDING

KNOX GROUP

ERIK RONNBERG

Contents

Part Three

"This is the General S." — replied Captain Coffin. — "What ship is that?"

A long pause ensued, and a number of voices were distinctly heard in earnest conversation on board the strange vessel.

"What ship is that?" asked Captain Coffin again.

"The Ark of Blood," cried the other captain, "commanded by the Demon of the Waters."

— William Comstock,
Voyage to the Pacific . . .

Part One

Prologue

The jagged tip of an ancient volcano, Easter Island rises like a beacon out of the vast, empty Pacific. It must have seemed exactly that to Hotu Matua, the mythic Polynesian discoverer who landed there sixteen centuries ago. He'd come with his extended family on an exhausting canoe voyage of thousands of miles, and with him he brought all the cuttings and seeds he'd need to begin agriculture anew. How he discovered the island, why he came from wherever he came, and how he survived the voyage are all parts of the essential mystery of the place. He was fortunate to find an unpopulated paradise that was well watered and heavily forested. The plants he'd brought flourished in the rich volcanic soil, and so did his people. Over generations, they multiplied and broke into clans, until their number approached 10,000 souls — far more than the island's sixty-four square miles could support.

This overpopulation initiated a grim descent into Malthusian inevitability. The clans began warring over precious resources, all the while depleting these resources even more in their obsessive construction of moai — the giant, enigmatic stone figures that have come to characterize the island. As conditions worsened, the people felt even more compelled to create the moai, as if these were their only hope of salvation.

In 1822, when the Nantucket whaleship Foster put in there, the condition of civilization on Easter Island was terminal. Vegetation and soil were depleted. Crops frequently failed, and there was no wood to build the voyaging canoes that would have allowed the islanders to escape. Construction of the moai had ceased, and the starving natives, who at that time "were 5 or 6,000 in number," were locked in a state of perpetual warfare.[1] Although it was one of the truly exotic mystery spots of the Pacific, the desolate, war-torn place could not have seemed very welcoming.

It was here that a young crewman approached Captain Shubael Chase with a bizarre request. The lad's name was Samuel Comstock. Though still in his teens, he was already the veteran of several merchant voyages and one whaling cruise. He came from a respected Nantucket family and was, by all accounts, an energetic and able sailor. His next whaling voyages would almost certainly see him promoted to boatsteerer, then mate, then captain.

But Samuel Comstock had no interest in the standard career path. While Captain Chase traded seeds to the natives for a few potatoes, Comstock approached and requested permission to be left ashore — to be marooned, to all intents and purposes — on Easter Island.

He was a strange young man.

chapter one

Jay Small's Big Hit

Vevay, Indiana, 1978

 In 1978, a midwestern book scout made a deal that would see him through the rest of his days.

Jay Small was one of that odd, itinerant breed who made their livings haunting estate sales, thrift shops, and old ladies' attics in hopes of finding rare and valuable texts. These characters flourished in the middle years of the 20th century, when we were becoming "modern" enough, most of us, to see no value in dusty relics, and rich and smart enough, a few of us, to collect these same objects with passion and vigor. In the rare-book hierarchy, Small was the man on the ground.

He specialized in Americana, and he

operated out of Indianapolis, Indiana. His beat was the central part of the Midwest — Indiana, Illinois, Ohio, Missouri, and Kentucky — territory that had been a conduit for American westward expansion. As people had passed through, they'd left books, pamphlets, and documents behind, a record of their activities and aspirations. The choicest of his selections might move along to bigger dealers in New York. These men would peddle their wares, by now substantially marked up and with an ever-increasing accompaniment of research and documentation, to more prestigious collectors and institutions.

Small had a friend and colleague who delighted in collecting and selling the works of the most obscure authors possible for a dealer in Indiana — Russian dissidents and Victorian English poets. He was woefully ignorant of Americana. So, in the somewhat perverse way these matters often proceed, it was he who stumbled onto the richest trove of Americana either man had ever seen.

Down in the southern part of the state, in a small town named Vevay, a mysterious collector had filled two buildings full of books and antiques. When the collector died, his family realized he had left them with a major logistical problem. To solve at least part of it, they were offering to sell all the books in one whack for $15,000. Simply by being in the right place at the right time, Small's friend got an exclusive on the deal, and Small talked himself into the arrangement, fifty-fifty. As was almost inevitable, the friend, lacking his own specialized knowledge, got the worst of the split. Jay Small lived off his share for another fifteen years. If anyone knew the identity of that original genius of a collector in Vevay, the information got lost over time.

Toward the end of his days, Small befriended a younger man named John Mullins. Employed by the sanitation department in Indianapolis, Mullins found the back of a garbage truck to be an ideal perch for an apprentice book scout. He was soon bringing his trash-barrel finds to Small who, in the course of researching and evaluating them, would instruct his pupil. Mullins operated without the benefit of a college degree, but higher education has never been a requisite for good book scouts. Some of the great ones were practically illiterate. What was required — and these were qualities John Mullins possessed in abundance — was a broad view of the world and an imagination for the place of books in it.

In fact, in this capacity, the pupil eventually outstripped his teacher. Small was a knowledgeable man and he possessed an excellent reference library, but in a way he was inhibited by his knowledge. For Jay Small, a book was of no importance if he could not find it in one of his references. John Mullins instinctively knew that a book might be of greater importance if the references had *not* discovered it. This principle applied even more to manuscript material. Each item was unique and quintessentially of its own time and place. For just this reason, Small didn't trust any of it. For just this reason, all of it fascinated Mullins.

It so happened that one of the items Small passed along to John Mullins at the end of his life was a manuscript book of some sort. Small had never sold it, because he'd never taken the trouble to decipher its contents. It was a notebook, a typical 19th-century product, bound in stiff cardboard covered with marbled paper and a leather spine, measuring perhaps eight inches by six. Inside there were about 150 pages of writing, all in the same hand, more crabbed in some places than others, but capable of being read if you were patient and had a good light.

Mullins discovered that it had been written by a sailor aboard a naval vessel, the *Dolphin*, in the Pacific in 1825. The author was a midshipman named Augustus Strong. Mullins knew a little about log-books. They tended for the most part to be perfunctory, dedicated to recording the necessary but tedious evolutions of ship-board life. This manuscript was something else entirely. It teemed with adventure — battles with savages, wrecked boats, and narrow escapes. It doubled back on itself and repeated a section twice in slightly different wording. It ended with a rescue of some sort.

To Mullins, in the Midwest, this manuscript at first seemed preposterous. How could so much have happened in the space of a few months? Was it an attempt at a novel? He went to his reference books. There he learned that Augustus Strong's ship had been sent in search of the survivors of the *Globe* mutiny. The mutiny and its aftermath were famous in the annals of American maritime history. After murdering the captain, three mates, and a black steward, Samuel Comstock and his three accomplices had forced their ship-

24

mates to sail to a remote Pacific island, where they intended to burn the ship and live among the natives in the manner of the *Bounty* mutineers. Fortunately, six of the honest crewmen, including Comstock's sixteen-year-old brother, were able to sneak back to the ship and sail her to the coast of South America. Eventually, word of the mutiny got back to the United States. Augustus Strong's ship, the *Dolphin*, was dispatched by the secretary of the navy to apprehend the mutineers and rescue the innocent crew members.

The manuscript Mullins had inherited from Small was a record of the voyage across the Pacific to rescue those survivors, and of the rescue itself. It also contained an account of the mutiny. It wasn't an attempt at a novel.

This was all well and good — in fact it promised to be quite good for Mullins. He realized that he was the owner of a potentially valuable item. But at this point his imagination failed him. Who, exactly, would be interested in such a thing? The question hung fire for a time, as such questions will. Then his instincts took over and served him well. He knew the manuscript would not become any less valuable with age, so he tucked it away until the proper opportunity presented itself.

And, after several years, the opportunity did indeed present itself, in the person of an antiquarian book dealer from Ohio who had known both Small and Mullins. His territory was the same as theirs, and his specialty, military history, overlapped the Americana field. It was a mutually beneficial relationship. This book dealer provided knowledge, contacts, and cash; Mullins and Small provided books to be moved up the food chain. On one of this dealer's visits to Indianapolis, Mullins showed him the journal. The dealer knew exactly what to do with it.

He called me.

He called me because we'd done business before and because he knew my checks didn't bounce and that I generally did what I said I was going to do. He also knew that, for the past twenty-five years, I had specialized in buying and selling rare books and manuscripts relating to America's maritime history. If anyone ought to know what to make of that sailor's journal, it'd be me.

Forty-eight hours later, the book was sitting on my desk, its layers of protective wrapping piled beside it. It was indeed a crew member's account of the *Dolphin*'s rescue voyage, and it contained a narration of what the survivors had told the men aboard the *Dolphin* about Samuel Comstock's mutiny and about their twenty-two-month stay on Mili Atoll. The manuscript was unknown and unrecorded in the literature of the *Globe* mutiny — not just rare, unique.

Singularity alone didn't account for the journal's enormous charisma. I can think of no other word for how that book struck me then, sitting there in its own glow, having miraculously surfaced, like some ghost ship, with its own long-forgotten tale to tell. Nor was this merely a bleached-out hulk, worn smooth by the ages. As I leafed through the manuscript I could hear Augustus Strong's voice relating each day's events, and it was apparent that, wherever Strong was from, this was the most exciting adventure he'd ever been on.

chapter two

Quixote Bent

Nantucket, 1819

 Samuel B. Comstock was descended on his father's side from respectable Rhode Island Quaker stock, whose number included Indian fighters and a state Supreme Court justice. His father, Nathan, came over to Nantucket as a young man to teach at the recently established Quaker Meeting School.[1] There he met an island woman named Elizabeth Emmet. Her father had attained local fame for his literary achievements and was deemed "an eccentric wight" by the family chronicler, William Comstock, who claimed, "He shunned society with all his might and main." Nathan, the young schoolteacher, won the approval of Mr. Emmet, the shy eccentric, and in

1801 Mr. Emmet gave his consent for Elizabeth and Nathan to wed. Samuel, their first child, was born in 1802.

In the sixteen years remaining until Elizabeth's death, the couple would go on to have seven more children.[2] Their second son, William, born in 1804, is of importance to the history of the *Globe* mutiny because he had literary ambitions. In 1838, he wrote a whaling novel entitled *Voyage to the Pacific . . .*, which included a character based on Samuel and provided valuable information about American whaling culture. Then, in a book published in 1840, he recounted his brother's life and his crimes aboard the whaleship *Globe*. This work combined lurid sensationalism, potboiling romance, and historical fact in a manner calculated to appeal to the popular tastes of its day.

A third son, George, born in 1808, also had an effect on the recording of Samuel's biography. As it happened, he accompanied Samuel on the *Globe*'s fateful voyage and witnessed and survived the mutiny. He was a primary, if not a proprietary, source for brother William, and his presence in the chain of information gives William's narrative a validity it would not otherwise possess.

★ ★ ★

According to William Comstock, Samuel was an engaging boy — intelligent, quick-witted, and absolutely ungovernable. There was a charming side to his personality, but it seems to have developed primarily as a means of extricating himself from trouble. The trouble invariably stemmed from his refusal to accept strict Quaker standards of behavior.

When young Samuel got thrown out of Meeting for bad behavior, he explained to his father that he'd only been laughing because the fat old man next to him had fallen asleep and slid off his bench. He was surprised everyone hadn't noticed, because the fat man's hat had fallen on the floor too, and had made quite a noise. Taken aback by this story, Nathan withheld punishment. Next day, he called the culprit before him. He'd checked with a friend, who'd been sitting right there, and he hadn't seen anyone fall asleep, or slide off a bench, or drop his hat. "No wonder," replied Samuel without missing a beat, "for he was asleep too." Samuel escaped a beating.

When he was six, they packed him off to Nine Partners, a famous old Quaker school outside of Poughkeepsie, New York.

This school provided a "guarded" education, which in Quaker terms meant one that adhered to the moral assumptions and practices of the Society of Friends.[3] Predictably, Samuel rebelled against this regime. As the headmaster prepared a switch to whip him for some infraction, Samuel said, "Ah, friend Mark, it will be of no use; Father has used up a whole poplar tree on me, already; but to no purpose." It's a beguiling anecdote, and a revealing one as well.

Fathers like Nathan have been dealing with sons like Samuel since time began. Such men cherish the spirit in their boys, because it is the energy of valor, of leadership, of accomplishment. They hope not to extinguish this spirit, even while they whip the boys — and their sons always get whipped, in one way or another — because, if given their head, the youngsters would run roughshod over the family, ultimately bringing harm to themselves and everyone else. The Nathans of the world become the first voice of society. They take responsibility for making it known that there are rules governing how people live together, and penalties for breaking these rules. That Samuel was exceptionally willful and Quaker standards exceptionally strict, only exacerbated the situation. Transgression and punishment were constant family themes, and punishment was corporal.

Samuel returned from Nine Partners after about a year, sufficiently grown to be beyond his mother's control. He found a hatchet somewhere around the house and took it for a plaything. The barn received its share of abuse, fences were chopped up, and, not coincidentally, the large poplar tree in the front yard went down, victim of that little hatchet.

Along with his extraordinary willfulness, Samuel exhibited another unusual quality. Once, as he and his brother William fought over possession of a hammer, William suddenly released his grip and the claw flew into Samuel's face, badly splitting his lip. He endured the injury and the subsequent stitches without a murmur. His high tolerance of pain would later impress his shipmates. For young Samuel, it rendered the frequent beatings even more futile.

A year or two prior to the War of 1812, the Comstock family moved to New York City. Nathan, probably unable to support his growing family by teaching school, had already taken a position as cashier at the Pacific Bank in Nantucket. After moving to Manhattan, he set up as a commission merchant in lower Manhattan, near Fulton Street. New York City directories for the next decade show annual changes of address for the Comstock residence and business as Nathan struggled to establish himself.[4] The years of Samuel's adolescence must have been hectic ones for his entire family.

During this time, Samuel became completely obsessed with stories of military action and with fantasies of battle. In New York, he joined a gang of street toughs named the Downtowners, who passed their time battling the rival Corlears Hookers. In the manner of many troubled youths, Samuel kept his own hours and often came home late at night, bloody and bruised. He carried the scars of these early battles to his grave.

In hopes of ending his obsessive military fantasies, Samuel's father gave him a copy of *Don Quixote*. As William recalled in his biography, Samuel devoured the book. However, only the battle scenes made an impression, while the intended lesson — the folly of Quixote's chivalric fantasies — "sank like dregs from his sight."

After a short and unhappy stint at a Quaker school on Pearl Street in New

York, Samuel's parents sent him back upstate to Nine Partners, but he ran away and was returned home. Next, they tried a Quaker school on Long Island, but he did no better there. The problem was always one of conduct rather than intelligence. He did well enough at his studies and continued to manifest an attractive personality, but he had an unyielding opposition to Quaker discipline.

At the age of thirteen, Samuel ran away from home in the company of an Irish construction worker he'd met at the digging of Harlem Heights. He wanted to get to Philadelphia, find a ship, and go to sea, but he was picked up in Elizabeth, New Jersey, and sent back to New York. His father found him so intractable and determined, that he finally acceded to Samuel's wish and got him a berth on a Liverpool merchant ship, the *Edward*, commanded by Quaker captain Josiah Macy.

Four months later, Samuel was home from the sea, and now, in the throes of adolescence, he began his lasting dalliance with the fair sex. He abandoned the simple attire of the Quakers and began wearing modern, big-city clothes. The Quakers disapproved. "It might have been better for some of the little she-quakers if they had done the same," reported William, "for, unfortunately, he was but too successful with the amorous sex." Young Samuel had found another use for his native charm. However, the skirt-chasing put a drain on his finances, and he was forced to supplement his income by stealing from his father's store.

It must have been a worrisome few months for Nathan and Elizabeth. By 1817, Samuel was back at sea again. His brother William claims that he shipped on the *Beaver*, a renowned China trade vessel commanded by Richard Cleveland and owned by John Jacob Astor. If so, Samuel's voyage would have been an eventful one.

These were the years when Chile and Peru were locked in their revolutionary struggle with Spain. Though nominally on a voyage to Canton, the *Beaver* was actually bound for Chile with a shipment of arms for the rebels there. She put into Talcahuano on the southern coast of Chile for provisions and fell into the hands of Spanish royalists, who found her carrying $140,000 worth of munitions.[5]

After languishing in a Chilean jail for some months, the young gunrunner shipped on the Nantucket whaler *George*, which was homeward bound. This ship had departed in February 1816 for the Pacific, and returned in July of 1818 with 2,100 barrels of sperm oil.[6] William wrote that, during his time aboard her, Samuel "conceived a violent antipathy to whaleships and their officers — an antipathy which never left him. The shocking ignorance and gross vulgarity of the *George*'s crew, was a theme on which he often dwelt with uncontrollable disgust."

Mild variants of Samuel's attitude are common enough in the literature of the sea, suggested by Richard Henry Dana Jr. in *Two Years Before the Mast* and often discovered in the testimony of young whalemen who, by virtue of their education and social standing, found life in the forecastle repugnant. Such men, in language reminiscent of William Comstock's, spoke of their shipmates as "contaminating," "wicked," and "averse to my feelings."[7] Dana and the others eventually made their peace with life in the forecastle, but Samuel was unable to. That he was also unable to accept Quaker authority and standards of behavior must have made duty on a Nantucket Quaker whaleship nearly intolerable.

When the *George* docked at Nantucket in 1818, Samuel learned that in his absence his mother had died. She was thirty-five years old. He returned to New York and, with his father completely preoccupied, resumed his wild ways, frequenting a notorious Lombardy Street whorehouse. After some months of this, Samuel expressed a wish to go to sea again. His father, in true Nantucket form, insisted he go on a whaleship. Samuel, uncharacteristically, complied.

At the ripe age of seventeen, he shipped aboard the whaleship *Foster* in 1819 for a three-year voyage. Perhaps he was able to convince himself that this time his whaling experience would be different. Perhaps his troubles at home and at school made him uncertain about his chances for success as a merchant or tradesman. There was a perpetual labor shortage in the whale fishery during these years, and the smart son of a Nantucket Quaker would have been snapped up immediately. In addition, the *Foster* was owned by Paul Mitchell & Sons of Nantucket. Samuel's mother had been related to the Mitchells, and this connection meant Samuel would secure a good berth.

Samuel had always been wild, but in his late adolescence the wildness took on an increasingly romantic tinge. Part of this was the charm and guile, which he used to advantage with the ladies. But there were also stories, military epics, and yarns, of which his brother says he had an inexhaustible fund. "His mind ran entirely upon battles, dark deeds, and perilous adventures of every description." This romantic strain now manifested itself in an odd but characteristic decision.

Samuel B. Comstock was not given a middle name at birth. He added the *B* himself, about the time of his tour on the *Foster*. The fact that survivors of the mutiny, in their later testimony, occasionally referred to him as "Samuel B. Comstock," suggests that he insisted upon it. William says he adopted the middle initial in honor of Captain William Burrows of the American brig *Enterprise*. A naval hero in the War of 1812, Burrows died at the conclusion of a fierce and triumphant sea battle against the British brig *Boxer*. Certainly the story of his exploits was well known, but it is easy to imagine Samuel Comstock, son of a schoolteacher, grandson of a Nantucket intellectual, student of *Don Quixote*, poring over the detailed and highly colored accounts of this battle, which appeared in such contemporary periodicals as the *Analectic Magazine* or the *Portfolio*, or in Bailey's or Wilson's early biographical collections of American naval heroes.

The umbrage he took at "the shocking ignorance and gross vulgarity" of his fellow whalemen suggests a lifestyle sufficiently refined to include literature. Even if Samuel only read the newspapers and popular journals of his day, he would have been deeply immersed in the exploits of America's first naval heroes. This kind of "worldly" literature was not a part of the curriculum of his Quaker peers, for whom "oil and ignorance . . . were staple commodities." Its content was directly antithetical to the "Peace Testimony" — that central tenet of Quakerism, which forbade violence against other human beings no matter what the provocation.[8] And, while feeding his appetite for violence and battle, Samuel's reading also provided the materials for similar fantasies of South Seas adventure.

Captain Cook's voyages had been published in cheap editions in America since 1774.[9] David Porter's narrative of his Pacific cruise against the British was extremely popular in Comstock's day. As well as battle scenes from the War of 1812,

39

Porter's book contained detailed descriptions of life in the Marquesas, exotic customs and rites, and beautiful native women. An even more compelling source for Samuel would have been the saga of the mutiny on the *Bounty*. Published accounts of the mutiny first appeared in 1790, and the story was included in American anthologies as early as 1806.[10]

Some of the mutineers were found at Tahiti in 1791, but Fletcher Christian and the rest of his comrades had gone on to places unknown. As it happened, a Nantucket captain named Mayhew Folger discovered their whereabouts during a sealing voyage in 1808. Captain Folger landed his ship, the *Topaz*, on remote Pitcairn Island, where he met the sole survivor of the mutiny and learned his story. The mutineers had run the *Bounty* ashore, landed all her goods, and then burned her. Over the next few years, sickness and treachery took their toll, until only Alexander Smith and his several Tahitian wives survived.

Folger recorded all this in the ship's log, which returned with him to Nantucket. The tale traveled back with him as well, and was reported everywhere. Samuel Comstock could hardly have missed hearing about it.

Desertions in the Pacific were so frequent that they were an expected part of whaling. This testifies to the power of such stories as the *Bounty* saga, and to the arresting qualities of the islands themselves. In 1842, for example, a young whaleman deserted from the *Acushnet* at Nuku Hiva in the Marquesas. His name was Herman Melville, and his fictional account, *Typee* (published in 1846), was a bestseller in its day.

Even more relevant was the story of David Whippey. He was a Nantucket lad, roughly Comstock's contemporary, who had a lust for remote places. In 1819, he signed on the whaleship *Francis* but deserted in Ecuador, and eventually wound up on an English brig trading sandalwood in the Fijis. Whippey took a liking to the islands and, as Comstock had done at Easter Island, requested his discharge from the ship.[11] During his trading activities, he'd made friends with the fierce Fijian warriors, and now they gave him a chiefdom of his own on the island of Ovalau. When Charles Wilkes and the United States Exploring Expedition landed there in 1840, Whippey had become an esteemed personage, "a royal messenger, or Matticum Ambau, . . . much looked up to by the chiefs."[12] He acted as Wilkes's interpreter and guide, and he provides a perfect example of the realization of a fantasy quite common in its day — common enough to have been held by such diverse figures as Herman Melville and Samuel Comstock.

By his seventeenth year, it was becoming apparent even to Samuel that he was not fitting well into society. He may have thought he'd inherited his maternal grandfather's literary and antisocial tendencies. Battle lore and island dreams mixed and surged in his imagination. On board the *Foster*, he could be seen restlessly pacing the deck in the fiercest storms, without a jacket, furiously soliloquizing. William reports that Samuel "had long dwelt on a romantic design of spending his life amongst the savages on one of the Pacific isles. He wished to be the only white man on the island, and doubted not in a short time he should be able to get himself elected king."

Such was the history behind Samuel B. Comstock's wish to be left ashore on Easter Island. Captain Chase denied the request. Comstock brooded.

chapter three

The Island and the World

Nantucket, 1812

 By the time of the *Foster*'s Pacific whaling cruise in 1822, her home port of Nantucket had come to occupy a highly specialized niche in the world of maritime endeavors.

As hard use by the early settlers depleted Nantucket's timber stands and farmlands, her inhabitants increasingly took their livings from the sea. According to local tradition, a stray whale wandered into the harbor in the middle of the 17th century. "This excited the curiosity of the people. . . . They accordingly invented, and caused to be wrought for them a harpoon with which they attacked and killed the whale,"[1] thereby fixing their destiny. More likely,

43

they learned their trade gradually from fishermen on Martha's Vineyard and Long Island. In 1690, they hired a Cape Cod whaleman to settle on the island and pursue the business, presumably instructing the locals in its finer points. A tower was erected near the shore, from which a lookout was kept. When a whale was spotted, the islanders would row out and kill it, then tow it ashore. There the whale's thick coating of blubber would be stripped off, cut up, and cooked in large cauldrons, called *try pots,* to extract its oil. The oil would then be cooled, strained, and stored in casks.

Until it was replaced by kerosene toward the end of the 19th century, whale oil was a preferred illuminant throughout the civilized world. It burned cleaner than tallow or vegetable oils, and was relatively easy to extract. It was also in demand as a lubricant, and was used in the tanning of leather, the manufacture of soap and paints, and the preparation of wool. It had always been an important item in colonial trade,[2] and by the mid-1700s, whale oil was Nantucket's major export.

At the beginning of the 18th century, a new level of specialization had been introduced. Legend has it that, shortly before

1700, a whale washed up dead on the south side of the island. Prior to this, the predominant catch of the shore fishery was a species called right whales. The carcass of this washed-up whale was shaped differently. It had teeth instead of baleen, and its oil produced a brighter, cleaner flame. Furthermore, the upper portion of its huge head, or *case,* contained a reservoir of an even finer quality oil. This creature was the spermaceti whale — so called, it is said, because its case oil, when exposed to air, resembled seminal fluid.

Nantucket had long known of the sperm whale. However, it was not until 1712, when Christopher Hussey encountered a school of sperm whales off the Massachusetts coast, that Nantucketers recognized the market possibilities of this superior species. They embarked with a will on the sperm-whale fishery in local waters, determined to make it their own. Whale-catching thus evolved from happy accident to deliberate hunt.

This innovation meant that ships had to travel farther in search of their quarry. As the whales along Nantucket's shores became spooked or hunted out, the industrious islanders began making offshore cruises — progressively longer, and in progressively larger vessels. Typically, a thirty-ton sloop (perhaps fifty feet in length) would sail a few hundred miles into the Atlantic. When a crew caught a whale, they would strip the blubber at sea, pack it in barrels, and bring it back to Nantucket to be rendered into oil.

Increasing the fleet's mobility had a positive effect. The whaling business grew throughout the 18th century, and Nantucket's infrastructure expanded to accommodate it. The commercial harbor was moved from Madaket, at the end of the island, to a more secure location at the island's center. Here, wharves, processing facilities, and warehouses began to be built, as well as buildings housing the necessary service industries — smithies, cooperages, chandleries, outfitters, and the like. The town, consisting of several hundred two-story shingled houses, grew up the gentle incline behind the harbor. Overlooking all were Nantucket's iconic windmills, with ropewalks — long sheds where hemp line was manufactured — between them. By 1780, the town boasted three piers, each three hundred feet long. These piers, one writer noted with pride, "are built like those in Boston, of logs fetched from the continent, filled with stones, and covered with sand."[3]

As vessels became larger, the *try works* were moved from shore to ship, enabling the blubber to be rendered into oil at sea. Whaleships thus became factory ships. One thing, however, did not change. The ships were still manned, as much as possible, by island men, "held together by ties of family, religion, Nantucket origins, and of one motive: to extract whale oil as cheaply as possible from the world's oceans and sell it as dearly as possible in the world's markets."[4]

In 1772, a Nantucketer learned the secret of making candles from the spermaceti, which had been separated out from the head matter of the sperm whale.[5] These candles were far superior to their tallow counterparts, and they quickly became another successful whaling-based export. The fact that Nantucketers could now utilize their own catch to manufacture candles gave them additional leverage in the marketplace.

At the dawn of the Revolutionary War, Nantucket dominated the sperm-whale fishery. Between 1771 and 1775, she sent eighty-five vessels manned by 2,025 men to the "southern fishery," as it was called. Her ships sailed as far as the coasts of

Africa and Brazil, and returned 26,000 barrels of sperm oil — about two-thirds of the national catch.[6] Wellfleet and Dartmouth, her nearest competitors, barely returned 10,000 barrels between them.

Nantucket's increasingly industrial culture fostered specialization. The price of sperm oil might fluctuate, but the islanders had positioned themselves to produce this commodity more cheaply, and in vastly greater quantities, than anyone else. As long as commerce remained open, they prospered. However, when the Revolutionary War disrupted normal channels of ocean travel and trade, Nantucket's single-market economy suffered.

The Continental Congress's restrictions on imports and exports crippled the islanders, who depended on exports for cash and who received their sustenance by importing goods. Indeed, almost everything on Nantucket — from the logs of her wharves down to the very foundation stones of her tidy wooden houses — had to be "fetched from the continent."[7] As the war ground on, the people of Nantucket found themselves caught between the patriotic demands of America and the reality that Great Britain had control of the waters surrounding their island. Food and firewood were in short supply. The price of a barrel of flour reached an astronomical thirty dollars. At the beginning of the war, more than one hundred fifty ships filled her harbor; by the war's end, there were just a few rotting hulks.

Nantucket survived the Revolutionary War, but just barely.[8]

Some Nantucketers tried to recoup their fortunes by exporting the fishery to what they hoped were more favorable political and economic climates. For a brief period following the war, Nantucket whaling colonies existed in such locations as Dartmouth, Nova Scotia; Dunkirk, France; and Milford Haven, Wales. But the real future of sperm whaling still centered on Nantucket.

As shipping resumed, the islanders again began to feel the benefits of specialization. With their energies concentrated on the single task, they got back in the whaling business quickly, and again they prospered. In 1791, for the first time ever, six Nantucket vessels rounded Cape Horn in search of sperm whales.[9]

Growth was accompanied by its own problems. Islanders experienced periods of oversupply and collapsing prices. Also, as more and larger vessels put to sea, a labor shortage arose among the native white population, and shipowners were compelled to look to the mainland for crew. Free blacks were brought to Nantucket and lived in shanties in the southern part of town. Nantucket Indians had been among the original practitioners of shore whaling, and the industry was quick to recruit aboriginal islanders from Martha's Vineyard. These men might attain the rank of harpooner — just below the officers of the ship — and they would share, according to their station, in the proceeds of the voyage. Similar opportunities were available to African Americans. There was at least one whaleship whose crew was entirely composed of blacks.[10] It was not until the industry expanded and Nantucket lost its preeminence as a whaling port that manning the ships became less an island affair and more a numbers game.

During the War of 1812, Nantucket was once again pinched cruelly between Britain and America. Commerce shut down, and the islanders found themselves again dependent on the largesse of the British for passage of ships carrying their basic necessities. Adding to their woes, a tender belonging to the British frigate *Nymph* set herself up in the waters between Nantucket and Martha's Vineyard. Her officer and crew acted more like pirates than sailors, plundering every vessel they could catch.[11]

Matters were equally bad for Nantucket's sperm-whaling interests in the Pacific. This part of the world was in political turmoil. Chile and Peru, involved in overthrowing the colonial rule of Spain, had established their own navies and authorized privateers to harass the shipping of Spain and its allies. Many of these privateers were no more than opportunistic bandits who used their wars of liberation as an excuse to prey upon American whalers. It also happened that, in the War of 1812, armed English whaleships found an excuse to disrupt unarmed American shipping. Although the "onshore grounds" west of the South American coast were a rich resource, their exploitation was proving costly for American whalers. Something had to be done.

The remedy appeared in the person of a salty little rooster named David Porter. The son of a Revolutionary privateersman, Porter went to sea in 1796 at the age of sixteen, and he received his baptism of fire in an encounter with a British man-of-war. Two years later, he entered the navy, and by 1811 had attained command of the frigate *Essex.*

In 1813, on his own initiative, he undertook to sail the *Essex* around Cape Horn. Based on intelligence he received while provisioning at the Chilean port of Valparaiso, he began an epic sweep of the Pacific in which he virtually destroyed British shipping in those waters. He cleared the way in the Pacific for American whalemen, and he established American ties with rebel governments along the South American coast.[12]

Reports of his exploits circulated widely in the newspapers of the day, and the ingenious and energetic little captain became one of the great heroes of American naval history. Certainly, he was a hero to the whalemen of Nantucket. In the course of his year in the Pacific, he captured a dozen British vessels valued at $2,500,000.[13] He also found time to leave his mark on the island of Nuku Hiva in the Marquesas, renaming it Madison Island, interfering in local wars, and claiming that island as a territory of the United States.[14]

While the cruise was driven by Porter's native energy, its imperialist agenda was a mark of America's nascent foreign policy. Ports such as Valparaiso in Chile and

Callao in Peru, as well as those on Pacific islands such as Hawaii, were provisioning stops for merchant and whaling ships, and securing trading privileges and support for American vessels in such places was vitally important.

In the last months of 1814, commercial excitement began to build along America's eastern seaboard. For more than two years, the War of 1812 had strangled trade, and with the end of that conflict, pent-up demand for goods was making itself felt. Nathan Comstock moved his business from 13 Oliver Street to the more heavily traveled Bowery, and back on Nantucket, managing agents P & C Mitchell[15] responded to the excitement by arranging for the construction of a new whaleship.

Whale oil was the fuel of choice for the lamps of Europe and America, and British attacks on the American whaling fleet had severely reduced the supply. The Mitchell partners, knowing that sperm-whale oil would be much in demand at the war's end, hoped to position themselves to cash in on the bonanza.[16] They wanted a vessel that was state-of-the-art, conceived specifically to expand Nantucket's sperm-whale fishery deep into the Pacific, half a world away from her home port. No doubt with this in mind, they named their new whale-catching machine the *Globe*.

chapter four

The Birth of the Globe

North River, Massachusetts, 1815

 A hard winter and the hazards of war probably confined the Mitchell partners to Nantucket,[1] but they still would have been able to tend to preliminary matters, sending their canny, careful letters by sporadic packet vessels through the British blockade. They struck their deal, finally, with a mainland shipbuilder named Elisha Foster.[2] His shipyard was located on the North River, a winding tidal creek thirty miles southeast of Boston. Foster had built other vessels (such as Captain Folger's *Topaz*, discoverer of the fate of the *Bounty*) for Nantucket owners prior to 1812. During the war years, he'd been as idle, though not as traumatized, as his Nantucket clients.

Before the building started, someone representing the shipowners would have sat down with the builder to work out the details of the project. Paul's brother, Christopher Mitchell, and Christopher's sons, Seth and Charles, were all involved in the ownership of the *Globe*. But Christopher's thirty-year-old son-in-law, Gorham Coffin, was the most active of the Mitchell partners and the most likely man to speak with the builder. In February 1815, the moment word of the peace reached Nantucket, he would have taken a boat to Boston to see to matters of finance and insurance[3] and arrange for badly needed hemp and cooperage supplies to be shipped back to Nantucket. When it was time to treat with Elisha Foster, he'd board the North River packet out of Boston Harbor.

The hardworking sloop's regular route followed the coast southeast to the mouth of the North River.[4] The river, broad and marshy at this point, tended back north for three miles, then narrowed and headed due west as the land rose up around it. The packet would stop to unload cargo at this narrowing, make another stop at a drawbridge, and then meander with the current through pristine New England countryside. The winding stream, the generous, wide pastures rolling back from it, and the towering trees behind must have seemed to Coffin a far cry from the windblasted heath of Nantucket.

But the idyllic landscape, in its own improbable way, was a center of industry. The pastures through which the sloop seemed to sail were strewn with stacks of planks and with timbers and sections of trees of all shapes and sizes — a veritable open-air warehouse for the raw materials used by the dozen or more shipyards he passed in the course of his inland journey.

Each of the "shipyards" was no more than a field full of timber, a few oxen, a shed, and a slip or two descending to the river, but almost every one boasted a vessel in some stage of completion. There was even a fully built ship that had been trapped in the river during the war, finally being readied for sea. The smell of hot pitch would have filled the air, along with an occasional oath or burst of laughter, the ring of the caulker's mallet, and the clean, sharp knock of broadaxe on oak. Somehow, custom, waterpower, and the availability of good timber had crowded these few miles with shipyards. And, as on Nantucket, specialization engendered a certain efficiency. Every man knew his job. There was no time wasted training people to new tasks, or figuring solutions to new problems. It had all been done before, and repetition ensured that it was done well. Word got around, and people who had an interest in these matters knew that this was a good place to come for sound vessels built at a reasonable cost. North River ships were known around the world. In Coffin's time alone, hundreds of vessels had been launched there.[5]

Whaleships were likely to remain longer at sea, with greater intervals between repairs, than were merchant ships. Therefore, the communication between the Mitchell partners and Elisha Foster would concentrate on how to make a strong, dry, seaworthy, good-going vessel, less concerned with looks than with durability. She would need a keel, stem, sternpost, transom, and counter timbers of the best white oak; frames also, sided thirteen inches, being spaced six inches apart; keel, fourteen by fourteen inches; good quality oak for planking, bottom plank three inches, all fastened with locust, driven clean through, and wedged both ends; ceiling of oak; clear white pine for the decks; the whole to be salted with rock salt from plank sheer to light watermark between the frames before leaving the stocks[6] . . . and so on at great length, in order that there should be no doubt, in Foster's mind, of the high quality required for each of the tens of thousands of components of this vessel. At ninety-five feet on deck, three hundred tons (more or less) by carpenter's measurement, she was one of the first of a new and larger class of whaleships, and the owners wanted her built right.[7]

Despite the attention to specifications, there were no blueprints, no plans. The "design," if it were even represented physically for this particular vessel, was in the form of a builder's half-model — perhaps one carved years before and used, with some variations, to build several vessels.

Assuming a cost of forty or fifty dollars per ton,[8] this new whaleship was a major expenditure for her owners. Elisha Foster, on his part, would pay two dozen carpenters for the better part of a year. Yet men like Coffin and Foster were so attuned to the fine points of marine architecture that they could arrive at agreement on the final design of the vessel simply by discussion. Their eyes had been trained by a lifetime of observing a variety of hull forms, and they could communicate these points quite accurately with nothing more than reference to mutually known vessels, a lot of gesticulation, and perhaps a pencil sketch or two. Coffin might tell Foster he wanted his vessel shaped along the lines of the *Frances Ann* — referring to a ship of similar model that Foster had built before the war — only fuller at the bow and easier bilged, so she'd roll more when cutting in. Foster might agree and suggest a slightly deeper keel. By the time they'd finished their discussion, the design of the vessel, down to its most intricate details, would be firmly fixed in each man's mind.

Though the technology of wooden shipbuilding had evolved over thousands of years, it had changed little in the last few hundred.[9] The procedures followed by Elisha Foster had been imported from England in the 17th century and, with gradual modifications, passed down through generations of fathers and sons, masters and apprentices.

Elisha would have gone into his timber yard and surveyed the timbers there, selecting the ones he thought would do for the *Globe*. Then he'd gather a gang of workers and go into the woods to find the timbers that were still needed. The timing was perfect, for this was a job that could only be accomplished in the winter. Under Elisha's practiced eye, an oak with just the right curves would be selected and felled, and the appropriate pieces of the frame would be sawed from the branches and roots of the tree and hauled back to the yard for final shaping.

Any sailing ship is an assemblage of complex curves, and from colonial times the precise forms of these curves have been constructed and replicated by means of wooden patterns called *molds*. After her completion in 1810, the molds from the *Frances Ann* would have been saved. In 1815, the Fosters hauled such molds out and, after modifying them to match the changes specified by Coffin, used them to determine the sizes and dimensions of the *Globe*'s timbers.

On firm ground sloping down to the river's edge, the keel was laid.[10] Each component composing this seventy-eight-foot-long backbone was marked and carefully trimmed to just the right length and size by men using broadaxes. These timber components would be bolted together, the entire long assemblage resting on the blocks and supports from which the vessel would eventually be launched.

After the keel was laid, the stem was *scarfed* to it by structural members, which defined the profile of the bow. (A scarf joint runs diagonally across the timber. It is strong and simple to cut.) At the other end of the keel, the sternpost would be fixed by a *tenon,* which fit precisely into a *mortise* in the keel timber.

Onto this backbone the ribs, or *frames,* of the vessel would be added. These were assembled from smaller components known as *futtocks,* which curved to the pattern specified by the mold. The reason Foster searched his woodlots for timbers with just the right shapes was that the frames would be stronger if their curves followed the grain of the wood. Oak was, generally speaking, the toughest and most sinewy shipbuilding material available, but even oak was more likely to split along its grain. So shipbuilders tried to use the wood's natural curves when fabricating the vessel's components. The massive thirteen-inch frames of the *Globe* were made up of two layers of futtocks, with staggered joints.

The ship's frames changed from V-shaped near the bow, to uniform U-shapes for several dozen feet amidships, then to a wine-glass shape near the stern. The U-shaped frames crossed the keel at right angles and hence were called *square frames*. With its component parts joined together, each of the *Globe*'s square frames, curving out twelve feet and up twenty feet on each side, would weigh more than a ton. Motive power for lifting frames was the human back.

Construction on the *Globe* probably began in February 1815. There were some snowy days; certainly weeks' worth of raw, rainy ones. But if a man didn't work, he didn't get paid. The men who built her turned out six days a week, sunup to sundown, rain or shine. For niceties, there would have been the material satisfaction of seeing this wondrous thing take shape, the humor and bonding that attends strenuous group activity among males, and two grog breaks a day, "11 and 4." Pay was about six dollars per week.[11]

Toward the fore and after parts of the vessel, the curves of the frames became tighter. These frames were constructed in symmetrical halves on both sides of the stem and sternposts. The bow had to narrow to a point while maintaining maximum strength. The stern, meanwhile, had to rise abruptly, spreading out from the sternpost. This *transom stern* was constructed by means of staging and shoring, which held the heavy upright transom timbers in exactly the proper position while the adjoining members were added, the whole forming one rigid structural system at the back of the ship. The geometry of these curves was exceedingly complex. Elisha Foster kept it all in his head.

While the stern structure was being completed, the planking of the vessel would begin. Men with *adzes* (Imagine a heavy, sharpened hoe at the end of an axe handle, and you can get some idea of this implement. In the hands of a skilled worker, an adze could split a pencil line and produce a smooth, even finish.) *dubbed* the frames, flattening their outboard surfaces for each successive plank, or *strake*. A similar process took place on the inside of the vessel, except that these interior planks were called *ceiling*.

This was before steam machinery had found its way to shipyards along the North River, so planking would have been cut at one of the waterpowered sawmills along the river's tributary streams. Plank was *lined*, or cut to its finished shape, by means of a fiendish device known as the *pit saw*.

The pit saw was simply a ripsaw blade stretched across a wooden frame several feet long and wide enough to accommodate the widest timber from which a plank might be cut. One fellow stood on a staging on top, while another man stood down in the pit in a continual shower of sawdust as the two of them sawed away, the plank being moved across the pit as the cut progressed. There is little doubt which was the most desirable end of the operation.

The planking was fastened by means of long wooden dowels, originally called *treenails,* and then corrupted to *trunnels.* They were made of hard, durable locust, which grew in abundance locally, and they were fabricated by pounding a straight-grained length of wood through a sharp circular die. The trunnel, something over an inch in diameter, would then be driven into a hole that had been bored for it through plank, frame, and ceiling — a foot or more of green oak — 10,000 or even 20,000 times. Banging in the trunnels was easy; it was the boring of all those holes, by hand, with a primitive auger, that was the terrible part of the job. What made locust especially suitable for this purpose was its great resistance to rot and the fact that it expanded more than oak when wet, thus ensuring the tightest bond.

While the planking ascended the ship's sides, *stanchions* would be built up between the frames at deck level. These were the vertical timbers to which the bulwarks and rails — the walls of the ship — were attached.

The *Globe* had a *lower deck* and a *weather deck,* both planked with white pine. The transverse deck beams, which supported them, were caught at each end by hackmatack knees bolted to the inside of the hull. A resinous species also known as tamarack, hackmatack was lighter and more durable than oak, and held iron fastenings better than any other wood.[12]

This gave the vessel three working areas. The lowest was the *hold,* that space at the bottom of the ship where oil and supplies were stored. The hold on the *Globe* was eleven feet deep. Above that, the lower deck was built, about five and a half feet high. This was where the officers and crew slept and ate, where blubber was processed, and where additional supplies were stored. Finally, there was the weather deck, open to the air, upon which the work of the vessel took place. Companionways, or stairways, leading below penetrated this deck. The forecastle companionway led to the crew's quarters in the forward part of the ship, and the cabin companionway led to the officers' quarters at the stern. In the center of the deck was a *main hatch* that led down to the *blubber room* where slabs of blubber were cut up for further processing. Just forward of the officers' quarters was a smaller space, called *steerage*, where the boatsteerers bunked. This was entered from the weather deck by an opening called a *booby hatch*. A special foundation for the *try works*, brick fireplaces where the blubber was rendered in large metal cauldrons, was constructed just aft of the foremast.

As streaks of plank ascended the vessel's sides, the caulkers would begin their work. The outer edges of each plank would have been beveled prior to fastening, and into this V-shaped slot between the edges of the planks, tar-soaked hemp strands known as *oakum* would be driven with a mallet and a caulking iron. The sound that mallet and iron produced, and its infinite repetitions up and down the thousands of feet of seams in the vessel, was the music of the shipyards. The caulked seam would be sealed, or *payed*, with pitch, and the entirety of the hull below the waterline would be sheathed with felt or boards and *coppered* — covered with a thin coat of copper sheets. The purpose of this coating was to prevent worms from eating the ship, and seaweed from slowing her down.

Laminated layers of timber secured with iron drift pins made up the last major component of the hull, the *rudder*. The part of the rudder closest to the ship might be a foot thick, tapering aft to its trailing edge. The rudder would be sheathed and coppered like the hull bottom and *shipped*, or hung, on the sternpost. The rudder stock projected up through the transom and the weather deck, where it was fitted with a heavy tiller to which the steering wheel was mounted.

Building this wooden ship involved a great number of complex and toilsome operations performed by two dozen men who walked at dawn each day through the pastures of Scituate and Marshfield to Elisha Foster's shipyard, worked until sundown in fair weather and foul, then returned home at dusk. Oak was heavy, every piece of it, and men of this era were not big.[13] However, the finished product stood as a testimonial to the art, intelligence, and perseverance of the men who created it. The whalemen who sailed in the

Globe trusted their lives to her, and she did not betray them.

A crowd was sure to gather on launching day. Gorham Coffin or other Mitchell partners were probably there. The carpenters would be on hand with their families, dressed in their Sunday best. Farmers, millers, and men from neighboring shipyards would drift across the fields as the moment neared, drawn by the spectacle as much as by the rum, lemonade, or cider provided to celebrate the occasion.[14] Workers would saw away at restraining timbers until gravity took control, sending the new vessel and her supporting cradle down the ways and into the river.

The *Globe*'s first cruise was a short one. Since she was almost as long as the river was wide, she'd have grounded on the opposite bank, amid the mess of flotsam that had once been her launching cradle and ways. But Elisha Foster and his North River pilot were ready for this eventuality. A massive hunk of timber known as a *kedge,* or *dead man,* had been buried in the bank a short distance downstream. Even while the ways still smoldered from the friction of the keel, Foster's men had a line from the kedge to the windlass on the bow, by which means they *warped* the *Globe* around into the current. Then, clumsy as a baby, she'd bumble her way downstream, unmasted and unballasted. From the mouth of the river, she'd be towed the few miles north to Scituate Harbor where she'd be masted and rigged.

The *Globe* was a ship-rigged vessel, and while the term "ship" in its modern usage might apply to any floating thing bigger than a boat, it had a very specific meaning in the days of sail. A true ship had three masts, all of which were *square rigged,* with the main driving sails laced to yards that lay square to the mast.[15] Traditionally, a spar, called a *bowsprit,* and its extension, known as a *jib boom,* projected beyond the bow of the vessel. The three masts (proceeding from the bow to the stern) were the *fore, main,* and *mizzen.* On the *Globe,* each of these masts was composed of four sections: *mainmast* at the bottom, *topmast* in the middle, *topgallant mast* above that, and *royal mast* uppermost. A horizontal spar, or *yard,* supported each square sail at its top edge from the top of each section of the mast. Thus, the *foretopsail* would be that sail hung from the second yard of the foremast. This basic scheme was complicated by the mizzenmast, which would be *gaff rigged* on its after side, with a sail called a *spanker* suspended by a gaff on top and a boom at the bottom. Additionally, at the bow, triangular sails known as *headsails* ran from the bowsprit to the foremast by means of supporting lines, or *stays*.

In practice, matters were considerably more complex. Dozens of additional sails and spars put in their appearance. The lines for supporting all these parts and for hauling them up and down numbered in the hundreds, each with its own particular name and function.[16]

So, while she lay in Scituate Harbor, a sophisticated and highly evolved wind-catching system was installed on the *Globe*. A new crew of men — the riggers — now swarmed over her, stepping her masts and supporting them by port and starboard lines, called *shrouds*, and fore and aft lines called stays. This was the *standing rigging*. Then the gear for operating the yards and sails, called the *running rigging*, was *rove off*. When completed, all of this presented a bewildering web of hemp rope (always called *line* when put to use aboard a vessel) and the blocks required to operate and adjust them.[17]

Coffin may have returned to Scituate toward the end, fretting over the thousand

71

details, urging the men as gently and continually as he could. If so, he would have been rewarded by a jubilant ride in the new vessel to Edgartown, or Old Town, as the place was known then. There, stores were rushed aboard the *Globe*, followed immediately by the crew, and by her first captain, an able man named George Washington Gardner.

Though her port of registry was Nantucket, the *Globe* was one of many whaling vessels that outfitted at Martha's Vineyard. The reason for this was an accident of hydrography that eventually spelled Nantucket's doom as a whaling port. There was a sandbar at the mouth of Nantucket Harbor that only allowed vessels of shallow draft to enter. As whaleships became deeper, they had to be loaded and unloaded outside this bar by smaller vessels called *lighters*. This added considerably to the expense and difficulty in receiving ships and sending them out. The simplest solution was to have these operations transferred to a more hospitable port, and Old Town was it.

On October 24, 1815, the *Globe*, under Captain George Washington Gardner, departed from Old Town on her maiden voyage to the Pacific. She returned from her cruise on New Year's Day, 1818, after twenty-six months at sea. It had been a record-breaking trip, "the first . . . bringing over 2,000 barrels"[18] of oil. Captain Gardner was rewarded for his efforts by being sent back out three months later. This time the *Globe* returned in May 1820, again bringing back more than 2,000 barrels. Gardner got another three-month break and began on his third trip on August 9, 1820.

Things were running like clockwork for the Mitchells.

chapter five

Comstock and Worth

 When the *Foster* returned to Edgartown in April 1822, Samuel Comstock had more on his mind than Easter Island or books of naval adventure. He'd been at sea for thirty-three months, and, as much as he may have detested the business of whaling, he'd completed a successful voyage. He was nearly twenty years of age, at what appeared to be the beginning of a respectable career, with money in his pocket and the whole world before him. Better still, it was spring. He bought a new coat and went to Nantucket for a visit.

He still had ties to island life. The Widow Eunice Mitchell, a relative of Samuel's mother, was also related to the

74

shipowning Mitchells. She had "lovely" daughters about Samuel's age. There were, possibly, a few Nantucket shipmates,[1] and there were old school friends such as Henry Glover, who would later be one of the few locals to consider Samuel with any sympathy.[2] The land breeze must have seemed sweet after all that time at sea, and the girls even sweeter. Brother William, who has little good to say about Nantucketers in general, remarks, "Their daughters are lovely beyond compare; their persons are graceful, and their eyes tell of an unexplored country, where everything which can enchant the soul, grows to perfection."[3] Samuel, the explorer of remote shores, tried his hand at the "unexplored country."

This bustling town of 8,000 people was as good a place as any to experiment with the delights of female companionship. Excursions were a favorite form of recreation for young Nantucket men and women; Siasconsett, at the end of the island, was the usual destination — probably because it was as far as one could get from the elders in town. Fishing parties were another amusement, and if New England custom held true, these would have ended with a landing on some remote beach, where a fire would be built, the baskets broken out, and a steaming kettle of chowder made from the day's catch.

Samuel's whaling experience would have been a great help with the ladies. "The dance is a favorite amusement of both sexes. But these dancing parties are very select, no youths being admitted except those who have 'struck a whale.' "[4] Another commentator tells how young fellows would assemble with the girls of the neighborhood, and simply sit and talk.[5] If one of them had recently returned from a whaling cruise, he was likely to be the featured speaker.

How charming these daughters must have seemed to Comstock, and how charmed their lives truly were at this age. Imagine being seventeen in a town where a third of the most eligible young men kept passing through in eager waves, having just returned from years at sea with hardened muscles, thrilling stories, and bulging wallets. A few months and they'd be off again, replaced by another lot.

While the courting must have been full of excitement, the reality of marriage was grim. "No sooner have they undergone this ceremony than they cease to appear so chearful and gay . . . the new wife follows in the trammels of Custom, which are as powerful as the tyranny of fashion . . . the new husband soon goes to sea, he leaves her to learn and exercise the new government."[6]

Samuel ultimately settled on a Methodist girl, "different from himself as can well be imagined . . . gentle, unassuming and pure in heart."[7] So thorough was his pursuit, that he feigned a conversion to Methodism. There was even talk of marriage.

On May 3, 1822, as the season got sweeter and Samuel became ever more deeply engrossed in his Nantucket explorations, the *Globe* sailed into port and tied up not far from the *Foster.* She was back from her third Pacific voyage, and it had been another good one.

They'd offloaded 2,025 barrels of sperm oil from her at Edgartown. Of course, the "barrel" was just a unit of measure, since they stored the stuff in wooden casks that were fabricated by the ship's cooper and were of various sizes. But it amounted to something like 64,000 gallons. 250 tons. Seventy-three whales cut in, and many more killed and lost — in the Newfoundland basin, the Azores, and off the coasts of Peru, Hawaii, and Japan.[8]

In those days, the thrifty shipowners did not hire longshoremen or cargo handlers. The crew remained aboard until all the work was done, sometimes living on the ship even while she was in port and performing maintenance that in later times would be assigned to riggers or carpenters.[9]

So, while Captain Gardner tallied his records and gabbed with the deputy customs officer, First Mate Thomas Worth would be pacing and sweating at the waist of the ship,[10] bawling down the main hatch into the gloomy hold, or hopping to the rail to ensure that each cask was handled with sufficient care.

They'd have gotten the two tiers of rider casks up first, as well as the *dunnage* — the oil-soaked cordwood that packed the load in place.[11] These smaller casks could be hoisted out full, using *can hooks* — specially shaped pieces of iron connected by a chain that grabbed the *chine*, or rim, at each end of the barrel. The barrels would be hauled up on a tackle, slung on the main yard, and deposited gingerly on the deck of the lighter that lay alongside. On days when the weather was kicking up somewhere "outside," little swells would insinuate themselves into Edgartown's elongated harbor, making the jobs of men like First Mate Thomas Worth all the more ticklish. If one of the people on deck lost his grip on the fall, or if the ship rolled, or the lighter bobbed at the wrong time, somebody could lose a hand.

Then they'd go to work on the ground-tier casks, four feet in diameter and stinking of bilge water. The oil in these monsters would be pumped out before the casks were removed from the hold. Those casks not containing oil — few in the *Globe*'s case — would have been filled with salt water, both to act as ballast for the ship and to keep the casks from drying out. On long voyages, particularly under the Pacific sun, the entire hold would be wet down several times a week to keep the casks swollen tight. In the off-loading process at the end of the voyage, even the largest casks would be emptied of their contents and removed from the hold.

It was tiresome, nasty work, particularly when it took place so close to home, and particularly when Thomas Worth's fiancée, Hannah Mayhew, was waiting on shore. They would be married this layover, and Worth must have been beside himself to bring his courtship to its proper conclusion.

★ ★ ★

Eventually, he did get home. Thomas Worth and Hannah Mayhew were married on July 4, 1822. He was twenty-nine years old; she was twenty-three. The newlyweds moved into the Worth family home on South Water Street in Edgartown, but they both knew this would not be for long. Thomas Worth was a company man, and he was on his way up.

He'd come from a countinghouse in Boston, perhaps scouted by Gorham Coffin out of one of the insurance companies in which he owned an interest.[12] He shipped as a seaman on the *Globe*'s first voyage, rising to second mate and first mate on her next two trips. A few more successful voyages would see him at the other end of Water Street — on Captain's Row.

In many ways he was a counterpoint to Samuel Comstock. They were two men of equal ability who seemed to be moving in opposite directions. Where Samuel resisted and struggled, Thomas had conformed, paid his dues, and succeeded. The discipline Samuel so hated came naturally to Thomas, a team player who had demonstrated his ability to take, and give, an order. The society against which Comstock had rebelled all his life seemed to be waiting for Worth with open arms.

It is also a fact that Thomas Worth had spent his entire seafaring career as a protégé of George Washington Gardner, who was recognized as one of Nantucket's most brilliant whaling captains.[13] In his career as commander of the *Globe*, Gardner had pioneered new whaling grounds and established a level of success unprecedented in the Nantucket fleet. He was a hardworking man of seemingly impeccable judgment who made his crew — and his owners — a lot of money.

There could not have been a better man with whom to study the intricacies of whaling, navigation, and command. Considering that the *Globe* had been at sea for all but six months of her seven-year career, there must have been ample opportunity for Worth to learn his lessons, and his steady advance indicates that he learned them well. In George Washington Gardner, Thomas Worth had the real-life mentor who, for Samuel B. Comstock, only existed in literature.

The *Globe* and her crew eventually followed the lighters from Martha's Vineyard to Nantucket where, relieved of her oil, she cleared the bar at the mouth of the harbor and worked around to the Mitchells' dock on South Wharf. If ships could sigh, she would have paused there with a sigh of relief. In the last seven years, she'd traveled 60,000 miles and brought back more than 6,000 barrels of sperm oil. Her average stay at home had been only three months. She'd worked hard and sailed well.

Now her slop chest and stores were inventoried, foodstuffs and personal gear were removed, and the records of her voyage delivered to the Mitchells. Unattached crew — transients from distant ports or men who'd been recruited to replace deserters on the voyage — were fastened upon by genial land sharks who'd do their best, over the next few weeks, to pick them clean. Nantucket men recognized friends and family in the crowd that surrounded the *Globe*, and were reunited with their loved ones, all excited and confused by the years yawning between them.

And, while it may be less romantic, the point of all this labor, danger, and separation was that the *Globe*'s cargo was also let off at Nantucket, where twenty candle factories — the Mitchells' plant first among them — waited with their own mercantile sort of eagerness for spermaceti, the finest part of the catch. Because of the cooling process involved in their manufacture, sperm candles were only produced between October and June. The reality was that the season was drawing to a close, and factories were vying for one last haul of case oil.

When completed, the candles would be shipped to ports throughout the United States, where their bright, clean flame made them something of a luxury item — the halogen lamps of their day. The common sperm oil was refined, cleaned, repackaged, and stored on Nantucket's wharves or in her warehouses, whence it would be shipped, ideally in seasons of peak demand, to supply the nation's lighthouses and to light her street and domestic lamps. Depending on market conditions, oil would also go to Europe or the Levant. Thus, whales killed off the coast of Japan supplied raw material that would be processed in Nantucket and might illuminate the waterfront in Marseille or the premiere of Beethoven's Ninth Symphony before the Philharmonic Society in London. Sperm whaling was truly an international industry, and in 1822, thanks to the genius of men like Gardner and Mitchell, Nantucket dominated it.

Whatever reality Samuel Comstock saw on Nantucket that May, it was insufficient to hold him. In one of the abrupt changes beginning to characterize his life, he left the Methodist girl (much to the dismay of her parents), the dances, and whatever friends he'd found, and returned to New York.

It is probable, though, that he was still on the island when the *Globe* tied up there. Samuel might have greeted some of her crew on Nantucket's sandy streets or sat yarning with them in the welcome warmth of an afternoon. He might even have been among the crowd gathered on the dock to greet the new arrival, and rubbed shoulders with Thomas Worth as he descended the gangway.

chapter six

The Brothers Comstock

New York, 1822

 At home in New York City, Samuel's family noticed a different side of him than had the whalers' daughters he'd so recently courted. He was off the *Foster*, and he swore to his brother "by all the saints on the calendar" that he'd never go whaling again. But his return was tinged with more than the simple joy of being among his family. This last voyage had changed him in some elusive but fundamental way.[1] His sixteen-year-old sister, Lucy, remarked that he had a "bad look" about him.[2] William, too, was surprised by his brother's appearance, commenting, "He had not grown much taller, if any; but his chest was broad and full, his limbs very large, his face and neck nearly as

85

black as an Indian's, while there was nothing left in his eyes of the frank and open expressions that once characterized them."

Regardless of what his siblings saw, Samuel returned full of resolve to make a go of life ashore. In an unprecedented effort to fit in, he asked his father to take him into the family business. The elder Comstock, of course, was delighted to do so. Nathan's shop, now on Front Street, was prospering, and after half a dozen moves, the family was comfortably ensconced in a house on Market Street in lower Man- hattan. If Samuel had suffered some obscure hurt on his whaling voyage, he was at least as intelligent as ever, and he still had his wit and charm. He'd be a welcome addition to the firm.

Very shortly, however, troubling patterns resumed. On the trip back from Nantucket, he'd romanced, and now announced that he was on the verge of becoming "engaged to," an English girl. While Samuel assisted in relocating the shop from 247 to 191 Front Street, he had a dalliance with a mulatto worker. He continued his obsessive fantasizing about military adventures and told one and all that he had added Burrows's name to his own.

To those steeped in the traditions of the Quaker faith, this must have been at least as shocking as his womanizing.

He and William saw a good bit of each other during this period. They were just a year and a half apart in age, and it is not hard to imagine them working and playing together. William reports Samuel's flirtations and experiences in brothels in a way that suggests he may have been more than a passive observer. Just as there seems something archetypal in the complex father–son dynamic of Nathan and Samuel, so the relationship of the two brothers assumes an iconic quality. In his biography of his older brother, William continually refers to Samuel as "our hero," even while acknowledging the horror of his crimes. Samuel was in fact a hero to his brother, a bold adventurer with ladies by the score and an endless supply of stories.

William relates some of these yarns as if they were gospel, and they give Samuel's life the flavor of a romance novel. In Valparaiso, for instance, during his tour on the *Foster*, Samuel met a dwarf prostitute and her rival, an "Indian Queen." He killed a black man who was trying to rob him, then fell in with a tough English sailor named John Oliver — a character who would reappear during Samuel's mutiny on the *Globe*. While such stories may seem the quaint stuff of melodrama, they also make it clear that Samuel was not cut out to be a merchant. The business bored him, and, while he was perfectly competent, he drifted away from it.

The shop and residence were located downtown, on the east side, near what is now South Street Seaport. In 1822, this area bristled with piers and wharves, and it was somewhere along this busy waterfront that Samuel's next adventure took place. William says:

> A patriot frigate lay at the wharf, and was nearly ready to sail. He contrived to ingratiate himself with her officers, and soon obtained an offer of a good position aboard of her. He gained his father's consent to go in the ship, and his every wish seemed gratified.

On its surface, it sounds as if Samuel was set to follow his idol, Burrows, into the navy. Such a reading casts Samuel B. Comstock in a noble light. This would be a patriotic channeling of his violent military fantasies, a successful attempt to find a compromise between his own nature and the world's demands.

But this was not the case. There were standard and consistently applied regulations in place at this time for the recruitment of officers by the U.S. Navy. A ship's wardroom gang did not grant warrants for prospective officers. They had to be approved by the president and by the secretary of the navy. It was impossible, in the 1820s, for Samuel to have been "invited" aboard a patriot frigate "as an officer."

Not since the Revolution had American naval forces been referred to as "patriot." However, there was another revolution then underway — several, actually — in countries along the western coast of South America, against the Spanish Royalists. According to William's account, his brother had already been involved on behalf of the South American "patriots" when he sailed on the *Beaver* in 1817. Speaking of this revolution William says, "The contest was at that time vigorously carried on between the Royalists and Patriots in South America." Almost every issue of American newspapers during these years carried the latest reports of the patriot struggle. Citizens of the young democracy felt a natural sympathy for their South American brethren, and our merchants were keenly interested in the disposition of ports along the western coast. If Samuel's unnamed vessel was in fact a South American ship being fitted out in New York to participate in revolutionary activities, the passage rings true both in terms of the realities of the day and the romantic bent of young Quixote.[3]

Unfortunately, the plan was doomed. William says that "some miserable Quaker bigots" convinced Nathan to retract his promise. "Our hero was informed [by his father] that he would receive no aid — not even the necessary outfits which, as an officer, he would need upon entering the Patriot Navy. . . . At that moment, the star of Samuel Comstock sunk in blackness forever!"

Those "Quaker bigots" were, in all likelihood, fellow merchants, shipowners, and sea captains upon whom, in the tightly networked community of Quaker businessmen, Nathan depended for his financial livelihood. They were expressing understandable concern about the violation of one of the central tenets of their pacifist faith.

Once again Samuel acquiesced. Here was a young man who had almost never done anything but what he'd wanted. His will seemed to brook no obstacle. From his earliest years he'd been running away, using his charm, lying, and even committing acts of physical violence in the service of his own ends. Yet, he did not sign aboard the patriot vessel.

Almost certainly, the fledgling officer would have had to bring a good deal of money with him in order to equip and support himself on that rebel ship. Just as certainly, Nathan would have been the only source for this. If he withheld funding from his son, the plan would have no chance of success.

Whatever the truth of the matter, Nathan's broken promise marked the beginning of the end of his oldest son's efforts to find a place for himself in the orderly world of the Quakers and their commerce. In the sequence of events leading up to the *Globe* mutiny, it is here that Samuel becomes an outcast.

Orthodox Quakerism was already lessening its hold on the social lives of its adherents (Nathan Comstock and his merchant friends may have emigrated to New York as part of a more liberal Quaker splinter group),[4] and in the coming decade, the religion would see the first of the serious schisms that led to its demise on Nantucket. In this respect, Samuel's rejection of Quaker religion and society was a course followed by thousands of people. However, the violent intensity with which he chose to pursue this course was of a different order.

In an attempt to escape the miasma, William signed up for a whaling voyage aboard the Nantucket whaleship *Governor Strong*.[5] As he was leaving, Samuel gave him a salty, if enigmatic, piece of advice. "Never get the lee side of a clue line," he said, "nor at the batter end of a jib downhaul!"[6] No doubt the utility of these words was proven once William was engrossed in the world of clew lines, batten ends, and downhauls, but he leaves his readers guessing as to whether Samuel's last words of advice had any deeper meaning.

Before William left to go whaling, Samuel shared one last ambition with his brother.

He told me it had long been a favorite scheme of his, to sail in some vessel bound to the Pacific Ocean, and while there kill the captain and officers, and take possession of the ship. He would then take the vessel to some island inhabited by savages, and after murdering all his crew, join with the natives, teach them the art of war, and raise himself to the dignity of a king.

After William's departure, in a reversal of his own, Samuel announced his intention to sign on another whaleship. Nathan, evincing classic fatherly cluelessness, was delighted. He promised his troubled son that if he made a success of this trip he would buy him a whaleship of his own to command.

chapter seven

Christopher Mitchell & Co.

Nantucket, 1822

This time, instead of simply turning the *Globe* around and sending her back out as they had in the past, her new owners, Christopher Mitchell & Co.,[1] found another chore for their workhorse. As it happened, they had a whaleship under construction. She was on the ways in Haddam, Connecticut, all framed and planked, but was stuck there now, unfinished. The master shipwright had taken sick, and the carpenters, needing steady work to survive, had moved on to other jobs. The partners knew they'd start losing money on their investment if they didn't get their new ship out to sea, so they emptied the *Globe*, took down her topmasts, loaded her with all the supplies nec-

94

essary to complete the new vessel's construction and rigging, and sailed her, like some lowly supply barge, over to Connecticut. It must have been a strange trip for the veteran ship and crew. Her shortened rig made for slow going and gave her the look of a vessel in distress. One merchantman actually hailed her and asked if she needed assistance.

At the shipyard in Haddam, they rounded up what carpenters they could and recruited the *Globe*'s officers and crew to complete the building gang. These men used the *Globe* for their dormitory that summer, and finished decking, coppering, and rigging the new ship. Then they coopered a supply of ground-tier casks, set them in her hold, filled them with river water for ballast, and sailed her back to Edgartown to be outfitted for her maiden voyage. She was called the *Maria*, and at 365 tons,[2] versus the *Globe*'s 293, she reflected the trend toward larger vessels and longer voyages.

As a reward for his years of excellent service, George Washington Gardner, the *Globe*'s old master, would assume command of the new vessel and be given part ownership in her. Because of his past success, it is likely that most of his old crew followed him to his new command, thereby relieving Christopher Mitchell & Co. of the problem of finding new men. One addition to the crew was Gardner's son, George Washington Jr., then about thirteen years of age. In an account written fifty years later,[3] George Jr. recollects that the building price of the *Maria* was $9,600. Gorham Coffin and the Mitchell partners had probably negotiated a good deal by completing the work themselves.

Then they brought the *Globe* back home and gave her the refitting she so richly deserved. Her old masts were unstepped and new spars fitted where necessary. New yards were crossed, and all the miles of hemp line that had become worn during her last two years were replaced with new standing rigging. Her anchor cables (in seven years the efficient Captain Gardner had never lost an anchor nor parted a cable)[4] were finally replaced. Her ironwork was checked and renewed; the ceiling in the hold was refastened; her white pine deck, gouged and scarred from chopping blubber, was repaired. Living quarters were aired and painted; her innards were flushed and pumped. They *heaved her down* — hauled her on her side and stripped her of her old bottom sheathing, renewed her caulking, pitched the seams, applied fresh sheathing, and coppered her anew. It was an undignified operation, but when it was done the old girl bobbed at her mooring like a debutante. Fresh black topside paint with white molding trim and a belt of *bright*, or oiled, wood from stem to stern completed her makeover. Then, in the autumn, as the day of her departure grew near, riggers bent new sails, rove off the running rigging, painted the lower masts, and slushed the topmasts and topgallant masts.[5]

All this while, the coopers had been *getting out* new casks. The biggest of these, four feet in diameter, would be stowed at the bottom of her hold, stem to stern, *bung up and bilge free* — with the plug at the top of each cask so it could easily be filled with oil. When the *Globe* was rigged for another voyage, they sailed her back to Edgartown. Here, these huge ground-tier casks would be filled with salted water to keep them from drying out and to act as ballast. Then, while the *Globe* rode comfortably by the dock, a seemingly endless stream of workmen tramped up and down her gangplank, loading her with all the supplies necessary to sustain her crew for two years or more.

Atop the ground-tier casks, they stacked more casks and filled them with fresh drinking water, 100 barrels of beef, and the same amount of pork, 10 tons of bread, 100 barrels of flour, and 1,500 gallons of molasses. Barrels of peas, beans, corn, dried apples, coffee, tea, chocolate, butter, lard, sugar, vinegar, rum, rudimentary spices, pickles, and cranberries rounded out the staples. Also down in the hold, they packed *shooks* — bundles of staves and hoops to make more casks — an extra suit of sails, spare duck to repair them, cordage for the rigging, and other dry goods. Along with these expendables came the tools of the trade, or *whalecraft* as the whalemen called them — the harpoons (200) and lances (three dozen) used to fasten and kill the whale; the dozens of spades and knives used in removing the blubber; the hooks, pikes, and chains used for hauling it about; and the skimmers, strainers, and dippers used in trying it out. They loaded trade goods, the cheap trinkets used in bartering for provisions, and slops for the ship's store. These included nearly every item the men would need in the course of the voyage — clothing, shoes, tobacco (nearly a ton), knives, and even pencils and paper. It was assumed every man would provide for his own needs, but inevitably things wore out or got used up or lost. Then came the heavy articles such as extra spars, try pots, and bricks for the try works. Finally, when sailing day was imminent, they piled aboard as many fresh vegetables as would keep, along with chickens and pigs and a little feed to sustain them until it was their turn to feed the crew.

Everything that would fit — that is, practically everything except the whaleboats, spars, and gear — was rolled up the gangplank in casks of various sizes. The heaviest casks would be grabbed at each end by can hooks and swung aboard by tackle slung from the main yard, the same way the old casks had come off. As the contents of these casks were expended or emptied, the containers would then be available for the storage of sperm oil. Thus arose the old adage that a whaleship left port just as full as she'd be on her return.

The whaleboats found their places *on the cranes* — suspended from davits over the ship's side. Some of the extra spars were lashed on a rack forward of the mizzenmast,[6] and others were lashed vertically to the main- and foremasts. But for safety and convenience, most of the stores went below. The casks whose contents were for immediate use would be kept in steerage or in the main hold *on the head* — end up, the rest to be hauled topside as needed.

It probably cost nearly as much to rig the *Globe* and ready her for sea as it had to build her.[7] Yet a decent trip would return a three- or four-fold profit on this expenditure. In her seven-year career, this fortunate ship had paid for herself many times over.

While the refitting went forward, Gorham Coffin and his partners at Christopher Mitchell & Co. started in on the critical business of rounding up a crew. The *Globe* was a three-boat ship, meaning she would launch as many as three whaleboats at a time. This was fairly typical for her era, though with voyages becoming ever longer, larger vessels were being used. In later years, four-boat and even five-boat whaleships became common.

Figuring on a crew of six for each of the three boats, plus three men left aboard as *idlers* to mind the ship, the *Globe* needed twenty-one men to do her business properly. A merchant ship of her size could have been managed perfectly well by twelve or fifteen men, but whaling was a more labor-intensive business. This may have lengthened the process of filling out a crew, but it also made whalers, in general, safer sailing ships than merchantmen. Many accidents in the merchant trade resulted from the ship being unable to shorten sail quickly enough when a sudden squall blew up. With the extra complement of men on a whaler, emergency operations could be performed that much faster.

Gorham Coffin was the point man for Christopher Mitchell & Co., while his father-in-law, Christopher Mitchell, and brothers-in-law, Seth and Charles, sat in stony silence, like Old Testament judges. Thomas Folger, the fifth managing owner, appears only as a signature on one of the ship's enrollment certificates. But when it came to recruiting men for a voyage, it is very likely that all five of the partners got busy, working their contacts through Nantucket society, to the Vineyard, to their agents on Cape Cod, and all along the New England coast. In this era, "connections" were the most vital element of the business. Nantucket's dominance was maintained by a web of consanguinity, friendship, and religion that extended throughout the region and into all of the related trades and professions.

They didn't have to look far for a captain. The recently wed First Mate Thomas Worth had spent his entire career — seven years and three voyages — aboard the old ship, and knew every inch of her. Gardner, the most successful captain in the Nantucket fleet, had been his tutor. As the *Globe*'s new captain, all Worth had to do was just what he and Gardner had done on their past three trips. He'd bring his owners "greasy luck" (they knew that the "luck" was mostly intelligence and skill), and, as a rookie captain, he'd come cheaper than his mentor had.[8]

Gorham Coffin, meanwhile, was starting another boy along the traditional career path. His brother had died at sea a dozen years before, leaving three children behind. One of these boys, Rowland, was now seventeen and ready to begin his days as a whaleman. Gorham signed him on this trip as assistant cooper. To serve as cooper, he recruited another Nantucket lad named Cyrus Hussey. Hussey already worked for Mitchell & Co., and, though he too was only seventeen, he'd proved himself sufficiently skilled to do the job. Cooper and assistant cooper were important billets aboard ship. The coopers fabricated the casks that would be needed during the voyage, but they were also responsible for executing all the boat work and general repairs. Now, Coffin had two of his own people in these slots. Furthermore, because of Thomas Worth's fealty, he could be sure that his nephew Rowland would be under the captain's protection.

Similarly, Thomas Worth's cousin had a son who was ready to try his hand at whaling. His name was Columbus Worth. At age fifteen he'd go into the forecastle as one of the greenest of the greenhands, but he'd provide another strand of loyalty. Worth also recruited two brothers from Tisbury on Martha's Vineyard, Stephen and Peter Kidder, ages eighteen and twenty-one. Being Vineyard men, they were known to and trusted by their captain.[9] Even better, they were experienced hands.

Worth and his Vineyard connection may have accounted for other recruits as well. By the time the *Globe* had gotten repaired and outfitted, it was late in the season, and most of Nantucket's talent was already at sea. It is probably no accident that all three of Worth's officers were from Martha's Vineyard. This was common enough at the time, and would have been perfectly acceptable to the partners at Christopher Mitchell & Co. Though personal rivalries between Nantucket and Martha's Vineyard had always existed, the two islands cooperated in the whaling industry to their mutual benefit. Many of the captains who lived in the fine houses on North Water Street in Edgartown had made their fortunes on Nantucket-owned vessels.

The manner in which these fortunes were acquired is one of the many fascinating aspects of the industry. Probably from medieval times[10] and certainly since America's earliest shore-whaling days, the fishery had been a cooperative venture. As early as the 17th century, the Dutch had formalized profit sharing into lays, and this method of distributing the proceeds of a voyage prevailed in American whaling throughout the 19th century.[11] Under this system, each man, in return for his share of the labor, would receive a certain fraction of the proceeds, while the owners would provide all the outfitting and supplies for a voyage. Lays, therefore, were always expressed fractionally. A captain might receive a lay of $1/18$, while a first mate might get $1/28$, a boatsteerer $1/70$, a foremast hand $1/150$; a cabin boy $1/250$. Although the ranges were fixed, there was nothing hard and fast about these numbers. A man's lay was negotiable, depending on his rank and his skill. Thus, an experienced and successful captain like George Washington Gardner might bargain himself into a short lay of $1/16$. The greener or less promising a man, the longer his lay was likely to be. The specifics of each contract were hammered out between the seaman and the owner at the time of signing. Where Stephen and Peter Kidder might talk the Mitchells into $1/130$, young Columbus Worth might have to settle for $1/180$.

While there were fortunes to be made in whaling, most of the money went to the owners and captains. If the *Globe* had a successful trip, she might gross something in excess of $50,000. From this amount, net proceeds were calculated by subtracting all the voyage's operating expenses as well as such administrative fees as insurance and commissions. Assuming a net of about $45,000, the seaman's $1/160$ lay might earn him $281.25 for his three years' work, but a couple of voyages could see him pulling a boatsteerer's $1/100$, or $450, while the captain's $1/16$ would bring him $2,812.50. After the captain and crew were paid, $30,000 or more would be left to the owners and investors. This was in the days when a very nice house could be bought for $1,000.

If she had a bad voyage — and there were plenty of those — things would not be so rosy. To keep himself in shirts, shoes, and jackknives, all purchased from the ship's slop chest, a sailor could easily find himself owing the ship fifty dollars. Stories abound of foremast hands who, at the end of three years of brutal and mortally dangerous labor, wind up owing more money than they made. Manifestly, the real way to prosper in whaling was to rise through the ranks, and in 1822 this was possible for anyone skillful enough to kill whales and lucky enough to survive. No doubt this was the future Gorham Coffin and Thomas Worth had in mind for Rowland Coffin and Columbus Worth.

In the midst of the turmoil of finishing and launching the *Maria*, and almost overlooked in the excitement of selecting new officers for the *Globe*, an experienced whaleman from a well-respected Nantucket family approached the firm through a Mitchell relative on Nantucket. The man had worked his way to boatsteerer on his last voyage, and one look at his solid, stocky frame and his bright, aggressive demeanor convinced the partners to offer him the same berth on the *Globe*. He had a fearless way about him, and would be an excellent man with the harpoon. In addition, he brought his brother George with him to serve as a greenhand before the mast. The boy was just turning fifteen, but would do well enough under his older sibling's eye. After he had been installed as captain, Thomas Worth interviewed the older brother and liked him immediately — in fact, he pronounced Samuel Comstock a capital fellow.[12]

Part Two

chapter eight

Irene

Gardner, Massachusetts, 1982

 Along with William Comstock's biography of his brother, there are two books that contain most of what is known about Samuel Comstock and the *Globe* mutiny. The first was written in 1828 by William Lay and Cyrus Hussey, survivors of the mutiny. Lay narrates most of the story, with Hussey adding details about events subsequent to the mutiny. Hiram Paulding, first lieutenant on Augustus Strong's ship, the *Dolphin*, wrote the second book, which was published three years later. The *Dolphin* was the vessel dispatched by the U.S. Navy to arrest the mutineers. These accounts have been sources for the dozens of articles and books that have kept the story of the mutiny

alive in American culture. As might be expected, the original works are scarce.

Another book from that era deals in a peripheral way with Samuel and the mutiny. This book had been my introduction to the brothers Comstock, and when I began thinking about the *Globe* affair, I dug out the copy I'd Xeroxed and read it again. It was a wonderful source, not so much for the factual information about Samuel Comstock as for the context it provided — what it had to say about Nantucket and whaling.

This book, also written by William Comstock, was entitled *A Voyage to the Pacific, Descriptive of the Customs, Usages and Sufferings on Board of Nantucket Whale-Ships*. It was published in 1838, fourteen years after the mutiny and two years prior to his biography of his brother Samuel. I first saw the book in Gardner, Massachusetts, in 1982, when I was learning my trade as a book scout. We were introduced by a woman named Irene.

Gardner was a hardscrabble blue-collar town in the 1980s, but earlier in the 20th century it had prospered as a center of the wooden furniture industry. Mount Wachusett, a little south of there, had been a popular tourist destination, and generations ago, the wooded countryside in that part of Massachusetts attracted wealthy summer residents. The whole area had been populated by a "better" class of people; people with an eye for the finer things in life; people, not to put too fine a point on it, with libraries. For years Irene had made her living extracting from attics and closets the books of these long-forgotten residents. In the course of her career she'd probably been in every old house in a fifty-mile radius.

Irene was talkative and large-boned, with orange hair pulled back in a bun and startlingly arched pencil lines where her eyebrows had once been. She and her husband, Walt, lived in a trailer that was raised up on cinder blocks. A wooden entryway and an "el" had been added, so that it resembled a deformed ranch house. She'd prettied up her front yard with plastic flamingos and wooden cut-out cows and ducks, and a clever windmill that, when the wind blew, made the little man on it seem to be sawing wood. There were concrete birdbaths on either side of the walk, but no birds ever came. Too scary.

Beside the house was a windowless cinder-block bunker about the size of a two-car garage. It may, in fact, have been a garage once, since it sported a wooden roll-up garage door. For the past thirty years, however, it had been where Walt lugged the books Irene had just bought. Each load would get put atop the last load, and there they'd sit until yet another load buried them. By the time I started shopping there, Irene's bunker was a solid mass of books, six or seven feet high, illuminated by three dim lightbulbs on the ceiling. The front door opened on a path that meandered through these books to a smaller door in the back corner. The way book scouts like me purchased our wares was to travel down this path, digging out interesting tomes and restacking the uninteresting ones on the other side of the path. The effect was similar to the manner in which a river cut its way through a valley. Over time, there was no part of that book heap the path hadn't scoured. Frequently, a fascinating vein would be discovered, producing an island or a new tributary. The pile had its hazards, though. One winter, Irene nearly died when a poorly braced section of the path caved in around her (books are very slippery in the cold), trapping her arms at her sides. She could feel the books draining the heat from her, and she knew it was just a matter of time before she succumbed. Fortunately, Walt came home from work and rescued her. A few winters after this episode, Irene gave me a piece of the birthday cake she'd baked for Walt. I took a bite, put it down, and lost it somewhere among those tens of thousands of books. The next April, I came upon it again, as perfectly preserved as the body of a frozen mountain climber.

Several times a year I'd drive out to Gardner and dig through Irene's books, emerging as filthy as a coal miner, with five or ten cartons of treasures. So it was, that in the spring of 1982, I found William Comstock's *Voyage to the Pacific* . . . . Initially, it attracted me not because of the *Globe* connection (of which I knew nothing at the time), but because I believed it was an overlooked source for Herman Melville's *Moby-Dick*.

Though obviously a work of fiction, Comstock's work contains an abundance of factual information about life on a whaleship, delivered with an authority that only could have come from experience. More fascinating to me were the echoes, throughout the book, of thematic elements of Melville's great novel. The similarities were almost eerie. A greenhand, like

Ishmael, has a comic debate with the owners about the lay he should receive. A black fool figure, like Pip, pleads for the life of a whale and then comes to an untimely end. A vengeful white whale, like Moby himself, smashes the whaleboats to flinders. Convinced that this seventy-two-page oddity had been a source for Melville, whose novel did not appear until thirteen years later, I began researching William Comstock. I found that literary scholars were largely unaware of Comstock's *Voyage*, which was good news for a book scout. But no documentation existed to prove that Melville had owned it or known of it, which was not so good. My theory was unsupported, except for one strand of evidence.

After their bout with the white whale, the narrator's ship is overtaken by a ghostly vessel. The captain calls out, "What ship is that?" and someone aboard her replies, "The Ark of Blood. . . . Commanded by the Demon of the Waters . . . like Lucifer, fallen from Glory and bound to Hell." It was, of course, Samuel Comstock — called "Samuel Hartwell" in this version — and the ghostly ship was the *Globe*, fresh from her mutiny. Something about the name of the ship rang a bell. I dug out my copy of *Moby-Dick* and found, in the "Extracts" portion at the beginning of the book, this quotation and attribution:

"If you make the least damn bit of noise," replied Samuel, "I will send you to hell."
Life of Samuel Comstock (*the mutineer*) *by his brother, William Comstock. Another Version of the Whale-ship* Globe *Narrative.*

Intent on following this trail to the end, I found and read William's second book, *The Life of Samuel Comstock*, and then the book by Lay and Hussey, *A Narrative of the Mutiny on Board the Whaleship* Globe, and finally, *Journal of a Cruise of the U.S. Schooner* Dolphin, the account of Lieutenant Paulding. I never did find conclusive proof that Comstock's *Voyage* had influenced Melville's *Moby-Dick*, but by the time Augustus Strong's manuscript journal of the *Dolphin*'s cruise landed on my desk, I was thoroughly versed in the Comstock family and the story of the *Globe* mutiny.

Everyone who has investigated this tale in a serious way, from Melville to James Michener (who wrote about the *Globe* saga in his *Rascals in Paradise*), has sat at a table with browned, fragile copies of the source books and mined them for the facts they contained.

Yet before the storytellers and historians could use them, these rare books had somehow to get recognized, and saved, and sent to someplace where people like Michener could access them. The job of gathering and preserving such artifacts is a very specialized niche in our culture, but because it provides our storytellers with their raw materials, it is a necessary one.

Despite her poor taste in lawn ornaments, Irene had sufficient imagination to collect and sell books that were worth saving. If it is an absurdly obvious truth that every book tells a story, it is also true that we take these stories too much for granted. Irene had an eye for the unusual, and because of her efforts, rare books were saved from the bonfire or the trash heap and ultimately found their way to libraries and museums.

That's where Irene's copy of William Comstock's *Voyage to the Pacific* went,[1] and that is where it sits today, waiting for the next scholarly detective to discover it and prove that Melville owned it, and that he used it as a source for Moby-Dick.[2]

chapter nine

William Lay

The Globe, 1822

 On December 15, 1822, the *Globe*'s topsails were loosed from the gaskets and left hanging in the gear. The mates hollered, and the men grunted, resplendent and stiff in their new sea clothes. Mooring lines were readied to be cast off, and the command was given to make sail. With far too much noise from aloft, fore and main topsail yards were hoisted, and topsails sheeted home. The fresh canvas shook out for a moment, then caught a breeze.

If the weather was fine, friends and relatives would have sailed out with the ship for a few hours, bringing hampers of food and forced good cheer, the excitement of departure all jumbled in with the melancholy of separation. If the day were windy and raw, as it was likely to be in the middle of December on Martha's Vineyard, the *Globe*'s departure would have been attended only by a disconsolate knot of mothers and wives, shivering at the end of the Coffin & Osborn wharf. In a flurry of sad looks from ship and shore, the deeply laden vessel would work her way out of Old Town Harbor on the tide.

All things considered, the Mitchell partners had been quite successful in recruiting a crew. They had the man they wanted as captain, and he brought with him two loyal foremast hands, the Kidder brothers, as well as his relative, Columbus Worth. Cyrus Hussey and Rowland Coffin, sound Nantucket lads, were minding the coopering duties. The boatsteerer, Comstock, and his little brother George were from a good Nantucket family, and the third mate, Nathaniel Fisher, had been born on the island. (The Mitchells knew Nathaniel's father, Amaziah.) That made for a core of reliable men from the quarterdeck down through the forecastle. Even the cabin boy, fourteen-year-old Joseph Prass, lived on Nantucket, though he was in fact Portuguese, having emigrated, as his people were just beginning to do, into the whaling trade from the Azores, or "Western Islands."

For his part, Captain Worth had been fortunate in finding a sufficient pool of experienced seamen on Martha's Vineyard from which to select his officers. First Mate William Beetle, Second Mate John Lumbert, and the other boatsteerer, Gilbert Smith, were all from the Vineyard and were known to the Worths, whose family connections spread through Edgartown and Tisbury. These connections had gotten the word out further, turning up Rowland Jones, a sixteen-year-old boy from a well-known seafaring family in Edgartown, and John Cleveland, twenty-one, from Tisbury. William Lay and Jeremiah Ingham had found their way to the *Globe* from Saybrook, Connecticut, and the last of the crew was filled out from a group of men probably sent over by shipping agents on the Boston packet. When the roster was completed, it looked like this:

CAPTAIN — Thomas Worth, Martha's Vineyard. Age 29

FIRST MATE — William Beetle, Martha's Vineyard. Age 26

SECOND MATE — John Lumbert, Martha's Vineyard. Age 25

THIRD MATE — Nathaniel Fisher, Martha's Vineyard. Age 20

BOATSTEERER — Samuel Comstock, Nantucket. Age 20

BOATSTEERER — Gilbert Smith, Martha's Vineyard. Age 20

COOPER — Cyrus Hussey, Nantucket. Age 17

ASSISTANT COOPER — Rowland Coffin, Nantucket. Age 17

COOK (and possibly STEWARD) — John Cleveland, Martha's Vineyard. Age 21

The foremast hands were:

George Comstock, New York. Age 14
Daniel Cook, Boston, Massachusetts. Age 22[1]
Holden Henman, Canton, Massachusetts. Age 23
Jeremiah Ingham, Saybrook, Connecticut. Age 17
Paul Jarrett, Barnstable, Massachusetts. Age 24
Rowland Jones, Martha's Vineyard. Age 16
Peter Kidder, Nantucket. Age 21[2]
Stephen Kidder, Martha's Vineyard. Age 18

122

William Lay, Saybrook, Connecticut. Age 17

Nahum McLurin, Bridgewater, Connecticut. Age 23[3]

Columbus Worth, Martha's Vineyard. Age 15

Cabin Boy — Joseph Prass, Nantucket. Age 14

In first-person narratives of this era, characters are introduced not as "Jeremiah Ingham," but as "Jeremiah Ingham of Saybrook, in Connecticut." For a young man a long way from home, that extra specificity carried a reassuring weight. Similarly, support of townsmen and neighbors could ease the psychological stresses of a long whaling voyage. Although the ship was a Nantucket vessel, only three crewmen — Coffin, Hussey, and Prass — actually lived on the island, while ten, including the captain and his officers, were from Martha's Vineyard.

More striking is how young this crew was. Although the captain and his first two mates seem marginally old enough for their positions of responsibility, most of the rest of the crew were mere schoolboys. In 1822, New England mothers sent their sons to kill whales in the Pacific Ocean at an age when modern parents would think twice about letting them have the car for a weekend.

In the culture of Nantucket and Martha's Vineyard, boys were raised with the expectation that they would become seamen. A writer in 1811 said of Nantucket, "Every child can tell *which way the wind blows,* and any old woman in the street, will talk of *cruising about, hailing an old messmate . . .* as familiarly as the captain of a whale ship."[4] Stories are told of young tykes harpooning family pets, and of ten-year-old boys going to sea. Younger bodies withstood the rigors of shipboard life, and young sailors tended to have less at stake on shore — though it is true that many seamen had families. George Washington Gardner, the *Globe*'s former captain, had a career that spanned thirty-seven years. During this time, he was home for a total of four years and eight months.[5] When a boy like George Gardner Jr. started taking up too much room around the house, off he'd go. Samuel Comstock went to sea at age thirteen, and his brother George at fourteen. Joseph Prass at fourteen. Columbus Worth at fifteen and Rowland Jones at sixteen.

It is also a fact that the quintessential

business of whaling was slaughter. As in the awful truth that young soldiers make the best killers, so young men made the best whale killers. They were easier to train and more likely to be in awe of authority. They had physical vitality and weren't old enough to realize that they wouldn't live forever. So the partners sat in their countinghouse like generals in their war rooms, sending the young men out to do their bloody business.

None of this was apparent to the boys who stepped aboard that December morning. William Lay, for example, had come down from Saybrook, Connecticut, probably in company with Jeremiah Ingham, "to go a-whaling." He may have been expecting tropical isles and wild adventures, but what he saw when he got to the top of the gangway was a frightening jumble of industrial clutter — the inscrutable mass of the try works; a grinding wheel; piles of canvas, rope, and chain; strange, cruel-looking hooks and spades; and tubs of potatoes, cabbages, onions, and other perishables strewn across the deck — all under a tent of ropes ascending skyward to the tips of masts as thick as factory chimneys. In his hometown at the mouth of the Connecticut River, he'd have seen plenty of schooners and packets, but nothing would have prepared him for the *Globe*.

The forecastle was a dim, crowded space, probably stinking of stale rum and vomit from those who'd come aboard drunk the night before. Lay's bunk was no more than a wooden shelf with a curtain in front, one in a wall of such shelves running two-high down both sides of the compartment. (If he sat up in the middle of the night, he'd bang his head.) The outfitter in Nantucket had already provided him with a bag full of straw — "a donkey's breakfast" — which, he now saw, just fit in the shelf and would serve as his mattress. This outfitter would also have provided him, at the enormous sum of sixty dollars,[6] a pine chest containing boots, pants, shirts and drawers, needles and thread, a sheath knife, buttons, a dipper, a dish, two bars of yellow soap, a jacket, and two blankets. By placing the blankets on his bunk, he'd be able to make enough room in the chest for the bag of things he'd brought from home, mostly extra clothes and a Bible. The chest would serve as his writing desk, parlor chair, and dining-room furniture. It would be given a spot among the other chests in the communal space, an open area about fifteen feet wide and ten feet long, tapering toward the bow and interrupted by the base of the foremast. The realization that this cramped, airless hole would be his place of comfort and solace, and that he would be sharing it with ten or twelve other people for the next three years, no doubt made his bowels churn.

Up on deck, Second Mate John Lumbert or Third Mate Nathaniel Fisher would have a lot to explain to the greenhands. William Lay would huddle with the likes of Rowland Jones, Columbus Worth, George Comstock, and Jeremiah Ingham while Mister Lumbert patiently rehearsed the sequence of tasks involved in leaving port. The seafaring terms, on first hearing, would have been mostly incomprehensible, requiring much repetition. Perhaps Lumbert would reassure them, telling them a seasoned hand would be beside them at all times. All they had to do was just what the fellow next to them was doing. "And be damned quick about it!"

The sudden change in tone would bring them up short, make them edgy and attentive, which, of course, was exactly where Lumbert wanted them.

It was a good thing, too, because just shy

of East Chop there was a sudden gust of wind and a loud crack followed by an aggravated flapping, curses, and a series of commands delivered with a volume and ferocity that William Lay had not thought possible. The bewildered greenhands would have been shoved aft by the mate, planted on each side of the mizzenmast, and commanded to haul away at the lines thrust into their hands, while, nimble as rats, four of the experienced men raced up the shrouds to furl the mizzen topsail.

More commands were given; the ship fell off the wind, turned, and began beating back the way she'd come. Much to their surprise, Lay and his mates found themselves back in Old Town Harbor before the afternoon was out because, they were told, the "crotchet yard had carried away." The cross-jack yard, as it was more properly known, spread the foot of the mizzen topsail, and, unaccountably, it had fractured before they'd even gotten into the Sound.

This would have been particularly galling to Captain Worth, since the vessel had just been rerigged. He opted to return and make his repairs ashore, which would avoid the necessity of expending spare supplies so close to home. More important, it would allow the Old Man a chance to let the shipyard gang hear about his busted spar, and perhaps one of those men might wind up wearing it, by God.

Back in Edgartown, the local men left the ship for one last unexpected night at home. The greenhands retreated to the forecastle, ears still ringing from the violence of the mate's commands. There, they'd spend the first of many nights, with only the cockroaches to entertain them.

That one night in port stretched to four. Probably the proper spar couldn't be found, or the riggers were gone, meaning the mates and greenhands had to accomplish the repair themselves. During this time, one of the crew, McLurin, disappeared — arrested or detained[7] — and was replaced by a nineteen-year-old sailor from Tisbury named Constant Lewis. Then, the men ashore would have to be rounded up, some of them doubtless suffering the ill effects of their premature liberty. When they finally did get off, the wind turned against them, and it was all they could do to beat around East Chop and put in at Holmes Hole on the northern extremity of the Vineyard.

Not until the twentieth of December did they get down Vineyard Sound. They sent the pilot home in his boat, and at last began to feel the long, regular heave of open ocean. The water turned metallic blue and seemed a heavier, denser fluid as it roiled around the ship. Greenhands and old hands alike stood along the starboard rail and watched the land disappear.

chapter ten

Comstock's Cold Eye

Christmas, 1820

 Stowed away at the Martha's Vineyard Historical Society in Edgartown is an oil painting of Thomas Worth that has somehow managed to survive the mischief of 175 years. The man looking out across two centuries is wearing a dark coat over a white shirt with a white cravat fastened down by an azure stickpin. A telescope, the sea captain's emblem, is tucked under his left arm. He has a lean, attractive face, with a sharp nose and chestnut hair of medium length combed forward on the sides in a dandyish way. He's looking to his left, not quite at us, but over our shoulders, his thin-lipped mouth slightly pursed, as if he were just about to tell everyone his thoughts on the

matter at hand. One has the sense that whatever Thomas Worth is about to say will be clearly explained and forcefully presented. We'd have no trouble hiring him to captain our ship, manage our business, teach our children.

Though unsigned,[1] the portrait is well rendered and professionally done — a society job, an artifact of the upper classes. The only anomalous thing about this portrait is the sitter's youth. Most of the sea captains who today gaze sternly down from museum walls are prosperous, ruddy, middle-aged men who waited until the prime of their careers — or even until they'd swallowed the anchor and settled down to comfortable lives ashore — before commissioning such a work. The fact that Thomas Worth sat for his portrait at the age of twenty-nine adds to the story the painting tells. The Worth clan was a large and prosperous one on Martha's Vineyard, and this capable young man was working his way up their exalted ranks. He'd just married Hannah Mayhew and just been promoted to captain of the *Globe*. His assurance is born of entitlement. As much to honor this expectation as to celebrate his marriage or his captaincy, his father commissioned the portrait some time before the *Globe* set sail in 1822.

This was the face William Lay and the other hands observed on that first day at sea when, just before the first dog watch, the captain called them all aft and addressed them from the quarterdeck. The sun had not yet darkened him. Months of indifferent food hadn't pinched his face, and the responsibilities of commanding a ship had not yet worn him down. His speech was formulaic, its content having been passed down through generations of seamen.[2] His manner of delivery gave the men an idea of what they might expect from their new master.

Flanked by Beetle, Lumbert, and Fisher, the new captain would have told his crew what he required of them — absolute obedience. They had a lot of work and a long voyage ahead, and the better they were at doing as they were told, the sooner they'd fill the ship with oil and come home. This meant following the orders of First Mate Beetle and Second Mate Lumbert and Third Mate Fisher. If these men issued a command, it was to be taken as if it came from his very mouth. There was only one reason they were out here, and that was to kill whales. Not to drink, not to play cards, not to fight, and not to pass the day sky-larking on deck. Greenhands would be given a week to learn their way about the ship, but every man was expected to do his duty at all times. When called from the forecastle, no matter what the time or weather, they were expected to jump up and be quick about it. The watches would be set immediately, and the boat crews chosen. There'd be men at the mastheads at all times, weather permitting, and the men on deck should keep a weather eye out, too. Five pounds of tobacco would go to the man who raised the first whale that got stowed down.

At the conclusion of this speech, Second Mate Lumbert and Third Mate Fisher, the *watch headers*, would have wasted no time in dividing the crew into starboard and larboard watches. While the ship cruised to the whaling grounds, these would be the primary divisions of labor. Each day was broken into five four-hour watches and two *dog watches* of two hours, during which the crew could eat supper. The second dog watch, 6 to 8 P.M., was the customary time of relaxation aboard ship. The men would stand *watch and watch*, starboard and larboard groups alternating their four-hour shifts. The dog watches also served to stagger the pattern, so the starboard gang would get the *mid watch* (midnight to 4 A.M., the least comfortable time to be awake) one night, and the larboard would stand it the next.

On the whaling grounds, a different system was adopted to allow for the fact that during the day all hands were employed on deck, ideally in catching whales, cutting them up, trying the oil out, and cleaning the ship for another round. During this part of the voyage, the crew would stand *boat's crews watches* throughout the night, each boatsteerer and his four oarsmen standing a four-hour watch, while the captain and mates slept. This would be the mode the men worked in for most of the voyage, and it necessitated the next order of business — the choosing of boat crews.

The *Globe* could lower as many as three boats when chasing whales. Each was commanded by a boatheader — Captain Worth, First Mate Beetle, or Second Mate Lumbert. Custom dictated that their boatsteerers, or harpooners, be assigned in order of seniority. Thus, Third Mate Fisher would be Captain Worth's boatsteerer, Smith would be Beetle's boatsteerer, and Comstock would be Lumbert's. The boatheaders then chose their four oarsmen, giving each of the three boats a crew of six. This would also be the division for standing watches on the whaling grounds, and it would ensure that each group of six spent a great deal of time working together. The boat crew thus became a social unit as well as a labor gang. There was competition between the crews, and there'd better be solidarity within the boat.

There was also competition among the boatheaders for the best men. The foremast hands would be lined up along the rail in order of their experience and examined carefully by the mates. "Like judges before a dog-bench," said one writer, describing a similar scene, "the mates strolled up and down the row, now feeling this man's ribs, now making that one bare his arm. . . . When the inspection had been finished, the drawing began. It was evident that the material had been studied carefully, for there was little hesitation and few words were spoken."[3]

Choosing the boat crews left three idlers from the *Globe*'s complement of twenty-one men. To these men would fall the responsibility of being shipkeepers. When the boats were out after whales, it was their duty to maintain communication with the boats by means of signals, generally flags flown from the rigging, or varying configurations of the sails.[4] Generally, the men whose duties required them to work all day, such as the cook and the cooper, were the idlers. Those who would be of more use aboard ship than pulling an oar, such as the ship's boy, fourteen-year-old Joseph Prass, might join them.

As Captain Worth promised in his address, two men were sent aloft to the fore and main royal mastheads[5] as lookouts, and the journey was considered officially to have begun. The lookouts would be relieved every two hours from now until the journey's end, as long as there was light enough to see by. A whale might turn up anywhere in the great ocean. There were even stories of vessels sailing into port, try works smoking, boiling out one last catch.[6]

As for the course Captain Worth set, Lay and Hussey's narrative, William Comstock's biography, and George Comstock's unpublished account provide only a general outline.

The logbooks from the *Globe*'s third voyage (1820–1822) and of the maiden voyage of the *Maria*, which sailed from Edgartown a month before Captain Worth departed in 1822, offer more detail.[7] Both voyages were under Captain George Washington Gardner, and the logs document his last trip in the *Globe* and his first in the *Maria*. Given the relationship between Worth and Gardner, it is likely Worth's voyage would have followed Gardner's general outline. Thus, these two logbooks give some hint of what Thomas Worth had in mind when he put to sea in the *Globe* in December 1822.

Under Gardner, the *Globe* had followed the Gulf Stream and prevailing winds from Edgartown to the Azores in the eastern North Atlantic, where she'd taken on more provisions, and sailed south past the Cape Verde Islands. Then she headed back across the Atlantic to the southern tip of South America, rounded Cape Horn, and sailed up the western coast before proceeding west along the equator, through a whale-rich area called the Offshore Grounds, toward Hawaii (known in those days as the Sandwich Islands). By the end of April 1821, she finished provisioning in Hawaii and departed Oahu for the Japan Grounds, where she remained until autumn. Then she returned to Hawaii, and after provisioning and recruiting, headed home with her 2,025 barrels of sperm oil. Gardner's next cruise, the maiden voyage of the *Maria*, essentially duplicated this pattern — Azores, Cape Horn, Offshore Grounds, Sandwich Islands, Japan Grounds, Sandwich Islands, Offshore Grounds, and home.

Not surprisingly, Thomas Worth followed the same route in the fifteen months of his voyage in the *Globe*. One difference is that he sailed from Edgartown four months later than Gardner had on the *Globe*'s prior trip, and perhaps for this reason, found the whaling somewhat slower on their accustomed grounds. Still, the overall pattern held. Depart December 1822; Azores January 1823; Cape Horn by March; Hawaii in May; Japan Grounds until the fall; return to Hawaii; cruise south and west along the equator.

The logs of the *Globe* and the *Maria* are replete with mention of other Nantucket ships sighted, of their captains, of how long they'd been out, of how much oil they carried. Despite the fact that American sperm whalers spread over thousands of miles and two oceans, they managed to maintain their communal bonds. Their vessels often *gammed*, passing time together on the open ocean, and sometimes *mated,* or hunted together, sharing the proceeds.

This was because the itineraries of these captains were well established. They weren't just sailing around out there hoping to bump into whales; they were following carefully developed routes based on close observation of the sea and its creatures, and on shared information. For this reason, it was quite likely that Nantucket whalers would turn up in the same places at the same times. Worth was headed for the Japan Grounds. If he hoped to catch whales there, he'd have to arrive in the proper season. And if he did, a good part of the Nantucket fleet was sure to be there with him.

The likelihood of such meetings resulted in some odd crossings. Two years earlier, in 1820, according to the logbook from the *Globe*'s third voyage, she had spoken the *Governor Strong,* and given her crew letters from home. This was the ship William Comstock probably sailed on when he went whaling, and on which he based the ship in his whaling novel. In his biography of Samuel, William claims to have met Captain Gardner in the Pacific, and this entry in the *Globe*'s log would substantiate his claim.[8]

More tantalizing is the fact that just a few days earlier, on Christmas, 1820, the *Globe* had spoken the *Foster*. This was Shubael Chase's ship, in the early part of her voyage to the Kingsmills and Easter Island. Like some conjunction of baleful planets in the winter sky, the *Globe* and First Mate Thomas Worth thus spoke William and Samuel's vessels within a few days of each other.

It is easy to imagine Samuel Comstock aboard the *Foster*, already full of detestation of whaling captains, and whalemen, and of whales themselves, standing apart from the men at the rail, giving the *Globe* a cold eye.

chapter eleven

The Greenhand's Education

North Atlantic, 1822

After the watches and boat crews were chosen and the look-outs sent up to the crosstrees, all hands were occupied setting the ship to rights. This would be the work of a week or more. The impulse behind it was not unlike moving into a new house and trying to find places to put things — shifting and stowing gear and rearranging fixtures to provide the most comfortable and functional environment. Those supplies that had been hastily piled aboard were stowed securely. Workspaces were cleared and arranged. This provided tasks for the greenhands, and an opportunity for them to learn their way around the ship.

There was also the job of tuning the rigging. The *Globe* had been built just after the War of 1812, and she was still held together aloft with hemp, wood, and leather. Though ironwork was used at places of greatest stress, this was well before the time of steel cable, and chain was a relatively scarce commodity. Iron would predominate over the next few decades, as sailing-ship technology evolved, but in the *Globe*'s era hemp had practical advantages. Whaling voyages were long. Things broke. The skill and ingenuity of able-bodied seamen was great, but there wasn't likely to be room aboard the *Globe* for a large stock of replacement iron. When hemp parted, any sailor could repair it.

So, in those first days at sea, yards were braced, and sails were reefed, furled, and shaken out, creating the necessary tension in the rigging. As the weather worked upon them, the lines of the standing rigging loosened and stretched, requiring constant maintenance. The crew's task of taking up the slack in shrouds, stays, and backstays would continue throughout the voyage.[1]

The sails that drove the ship and the spars that supported them were balanced

143

in a complex and dynamic system of opposing forces. Some sails lifted, while others depressed the vessel as they drove her forward, each sail affecting the flow of air upon the next. The masts were raked at slightly increasing angles, and the yards were corkscrewed in subtle ascending spirals. Captain Worth immediately began earning his keep by directing the mates and crew in bringing all these forces into their most efficient configuration, then altering them continually throughout the voyage to get the best trim in constantly changing weather conditions. The rigging was indeed made taut, but in a most calculated way, and as a unit.

The other major order of business — and this was as critical to the success of the voyage as sails and rigging — was preparing the whaleboats. Now that the boat crews had been chosen, each crew would work on readying its own boat, and thus would begin the process of forging six individuals into a team. These boats were hung from davits — wooden arms extending over the sides of the ship, from which the boats could be lowered and raised. On the *Globe*, and all other three-boat whaleships, there were two boats on the port side. The farthest one aft was called the *larboard boat* (the more ancient designation, *larboard*, rather than *port*, remained in use aboard whaleships to designate the left side of the vessel). The boat forward of the larboard boat was known as the *waist boat*. There was also a *starboard boat* on the right-hand side of the vessel, aft. The captain and his boatsteerer, Third Mate Fisher, would command the starboard boat. First Mate Beetle and Gilbert Smith would take the larboard boat, and Second Mate Lumbert would have Samuel Comstock as his boatsteerer in the waist boat.

These slim and graceful double-enders were perhaps the loveliest creation of the industry; they were adapted to their highly specialized task, and were aesthetically pleasing. Clifford Ashley calls the whale-boat "the best sea-boat that man could evolve, with no limitations to size, weight, or model."[2] In the *Globe*'s day they were about twenty-six feet in length, six feet in beam, and two feet deep, light and strong, with thin steamed oak ribs and a cedar *lapstrake* hull, where each piece of exterior planking was overlapped by the one above it. This type of construction added strength to the boats and perhaps some stability, for the whaleboats of this period were round-bottomed, with plank keels

145

and no rudder or centerboard.[3] They sometimes carried sails that would have been used for traveling long distances downwind, but they usually rowed to the whale, guided by a long steering oar.

The five whaleboat oars came in three lengths, ranging from the eighteen-foot *midship oar* to the fifteen-foot *after oar*. Each crewman pulled a single oar that extended across the boat and was held in place on the opposite gunwale by two tholepins. The arrangement was perfectly balanced — one long and two short oars extending from the starboard side, alternating with two medium oars from the port side. Each oar had its particular name and function.

In the *Globe*'s waist boat, for example, Comstock, the harpooner, sat to starboard on the bow thwart, pulling the *harpoon oar* fixed by tholepins on the port side. Aft of him on the other side sat the man who pulled the *bow oar* which extended over the starboard side. This job required both strength and experience. The bow oar (the man and the oar were called the same) was responsible for seeing that all the gear was clear of the line, and for tending to the wants of the man at the head of the boat — Comstock when harpooning the whale or

Lumbert when lancing it. To bring the boat up close to the whale so Lumbert could deliver his killing thrust, the crew would have to haul in the whale line. The bow oar brought the line aboard.

After the bow oar came the *midship oar*. This, being the power oar of the boat, required a man of strength. Constant Lewis, the big crewman picked up at the last minute in Edgartown, would have been a good candidate. Then came the *tub oar*, manned perhaps by the twenty-two-year-old mulatto Daniel Cook, and finally the *stroke oar*. This was one of the two shortest oars, and the lightest pull, and was usually handled by the smallest man in the boat, someone like George Comstock, Samuel's kid brother. Tub oar and stroke oar were responsible for the management of the line as it was hauled back into the boat. (Harpooner's and stroke oars were of equal length and were the shortest. Bow and tub oars were of equal, intermediate length, and the midship oar was the longest.)

For purposes of command, the number of oars on each side also identified the oarsmen. The two men on the larboard side of the boat were known collectively as *Twos*. The three oarsmen on the starboard side were called *Threes*. Thus, if the whale suddenly surfaced off to his right, Lumbert might command, "Pull Three!" and make the appropriate adjustment with his steering oar.

This talk of arrangement raises a peculiarity of the whaling business. Comstock, the harpooner, was officially known as the *boatsteerer* because after he harpooned the whale, he would move to the stern of the boat and exchange places with the *boatheader*, Lumbert, who would then move to the bow and *lance* the whale, delivering the killing thrusts. Comstock, now in the stern, would guide the boat during this phase of the operation and hence was called the boatsteerer, even though it was he who initially *planted* the harpoon that made the boat fast to the whale.

Precisely how this curious division of labor came about is lost to custom, but it must have evolved from the demands of use. The boatheader, as the whaleboat's officer, would always be in the position of command. He would be best suited to steer the boat in the chase, a backbreaking row that might take hours and require threats and humiliation, as well as supplication and outright cheerleading. Similarly, as the most experienced man in the boat, Second Mate Lumbert would be the logical one to thrust the killing lance into the whale's vital organs when it lay exhausted from its efforts to escape. If an emergency arose, and the whale line needed to be cut, the decision to do so would be Lumbert's.

Considering what went on in a whaleboat when all hands were doing what they were supposed to do, the amount of gear it carried was incredible. Along with its long oars and oarsmen, the boat carried five paddles used for silent running when the men, in flat, calm seas, made a final stealthy approach to the leviathan. The boat also carried wooden piggins for bailing and one for wetting the line, and a lantern keg containing tinderbox, matches, candles, pipes, tobacco, and hard bread. A signal lantern might save their lives if they were still at sea when darkness set in. A keg of fresh water would be stowed on the boat along with a flat, square board (known as a *drag,* or *drug,* or *drogue*) that could be bent onto the whale line in order to tire a frisky whale. Flags mounted on long poles (*waifs*) were used to mark the dead whales. These found their place at the bottom of the boat along with a boat hatchet, foghorn, and various knives, boat hooks, a grapnel, and assorted marine hardware. Canvas nippers were needed for handling the whale line, and a box compass for emergency navigation.[4]

The whale line was made of the finest hemp,[5] about three-fourths of an inch in diameter, loose-laid, soft and pliable, and only lightly tarred to preserve its flexibility and tensile strength. When a new coil was broken out, it might be run up through the rigging or towed behind the ship to remove all twists and kinks before being carefully coiled, Flemish style (a flat, tight spiral working in from the periphery), in the line tubs. This was a matter of life and death, not neatness or style. A good harpooner would be as careful in the operation of coiling it as a man packing a parachute.

Each whaleboat of this era carried a single large tub containing as much as 200 fathoms (1,200 feet) of line. The tubs were kept outside the boats until the crew was ready to launch. In use, the line was led aft from the tub; around a *loggerhead,* or post, at the stern; then forward to a notch in the stem called the *bow chock.* Twenty or thirty feet of line was fed through the chock, then coiled and placed in the *box* just aft of the chock. The end of this coil was tied to the harpoon.

The line was kept in place as it ran

through the chock by a *chock pin*. This was a whittled piece of oak, whale bone, or bamboo about three-sixteenths of an inch in diameter and four or five inches long. Back on Nantucket, a boatsteerer like Samuel Comstock might wear a chock pin in his lapel as a badge of honor. It signified that he had killed a whale, and, reportedly, it carried great weight with the ladies.

Whale line was never fastened to the boat; the friction of its passage around the loggerhead was sufficient to prevent it from flying out too quickly. If a strong whale exhausted the supply of line in one boat, it could be *bent* (tied) onto a fresh tub of line in another boat, assuming one was nearby. Otherwise, the bitter end was bent to the drogue, which would then be tossed overboard, marking — and tiring — the whale on the other end.

Finally came the harpoons, lances, and spades used in fastening to the whale and killing it. These whale-killing tools were known collectively as whalecraft. The harpoons used on the *Globe* were double-flued, like a classic Indian arrowhead, forge-welded onto a wrought-iron shank about a half of an inch in diameter and two feet long. They needed to be extremely malleable and ductile, because the action of a dying whale could bend them like corkscrews, and if they broke the whale would be lost. At the other end of the shank was a socket into which a six-foot hickory pole was attached. The whale line was securely fastened to the shank, and the socket's taper prevented it from being pulled out. The head of the *iron,* as whalemen called their harpoons, was incised with the name of the ship and some identification of the boat from which it came. So, the men of Lumbert's boat might engrave their irons "GWB" or some other unique designation, signifying that the iron was from the *Globe*'s waist boat. This let the crew know who struck which whale, and it also honored the ancient custom that "marked craft claims the fish." In theory at least, if two ships should dispute the ownership of a whale, the marks on the irons would settle it.

A good harpooner might be effective from as far as thirty feet. If the iron truly launched, the two-foot shaft would penetrate the blubber and fix in the flesh of the whale. The wooden shaft would then break away. Two harpoons were kept at the ready, supported by a Y-shaped *jack,* or *crotch,* near the boatsteerer's grasp. The second harpoon was attached to the line with a loose loop and plenty of slack. If the first effort missed, the second would follow in short order. Ideally, the whale might be *made fast* by both irons. Every boat would carry three or four spares, in addition to the two at the ready.

It has never been satisfactorily explained how men like Comstock and Lumbert were routinely able to change places over the length of the boat, dancing around all that gear and four exhausted oarsmen (and their oars); the whale line rushing out of the tub to the stern, then forward over the bow so fast it needed to be kept wet so as not to char (this was the tub oar's job); the boat moving through the ocean at ten or twelve miles per hour; the crew trying to stow their oars and bail, or simply hoping to avoid being plucked out of the boat by the rushing line. There are many stories of this happening to men. The wonder is that it didn't happen to them all.

After Comstock and Lumbert had exchanged places, and after the whale had run itself to exhaustion, the boat would creep up on the wounded beast. The bow oar and his mates would haul the line in until the boat was steady against the whale, at which time Lumbert would thrust the lance into the whale, *churning* it, probing for the animal's lungs or the large blood vessels under the lungs, referred to as the whale's *life*. The lance was a five- or six-foot iron shaft on a wooden handle of the same length. Its blade was petal-shaped and razor sharp. Application of this instrument was followed by a hasty retreat. If Lumbert had done his job properly, the blood would come up in thick red clots. The whale would go into its *flurry* and expire within a few minutes. The boat always carried spare lances, just in case.

Once the excitement was over, the men had a dead leviathan to deal with. At this juncture, they'd use another tool, the boat spade, to cut a hole in the whale's head, lip, or tail. They would then attach a line and begin towing it back to the ship, which, they ardently hoped, would be sailing toward them to secure the catch.

Each whaleboat weighed five or six hundred pounds. When the crew had assembled, sorted, and stowed all the requisite gear; sharpened the irons, lances, and spades; coiled the line in its tubs; loaded it all in the boat and climbed in themselves, the total weight of men and gear came to nearly a ton. A practiced crew could have their line tub in the boat and be down in the water in a minute or two.

★ ★ ★

That was the idea, at least. Those first days would see the ship tidied and tuned, greenhands oriented and educated, boats raised and lowered, novice crews forged into efficient killing teams. In actuality, that first week at sea was likely to be taken up with a different sort of activity. George Washington Gardner Jr. spent his earliest days aboard the *Maria* paralyzed with seasickness. "I thought I should dye," he writes. "Then the cook . . . brought a piece of fat pork about as big as a walnut and a string tied to it and molasses. He said, drop the pork into the molasses and swallow it and pull it up by the string two or three times, it would cure me. I did not try his remedy but the very thought of it clapped the climax and I give up all idea of living."[6]

Benjamin Ray, first mate of the *Maria*, summed the situation up in his logbook entry, six days into their voyage: "Employd fixing the rigging and boats we can do but little so many of the ships crue seasick and green."

chapter twelve

First Trial

New Year's Day, 1823

 The frightening new-
ness of the ship and
the misery of seasick-
ness had probably
just worn off when,
on January 1, 1823,
Captain Worth and his men experienced a
heavy gale from the northwest. In his narrative of the *Globe* mutiny, William Lay calls
the storm "our first trial." Twelve days of
sailing had put them literally in the middle
of the Atlantic Ocean, and the waves had
the reach of 1,500 miles behind them to
build, under the wind, to gigantic rollers.
They may have been in the Gulf Stream,
but it was still winter. Life expectancy for a
swimmer in those waters would be measured in minutes.

Such was the skill and experience of

Thomas Worth and his officers that the gale was regarded as a boon rather than a life-threatening situation. "As the ship scudded well, and the wind was fair," William Lay wrote, "she was kept before it, under a close reefed main top-sail and fore-sail, although during the gale, which lasted forty-eight hours, the sea frequently threatened to board us, which was prevented by the skillful management of the helm."

Scudding was an approved technique for riding before a storm. The main topsail was employed because the main course, the largest and lowest sail on the mainmast, would frequently be sheltered from the wind by the height of the seas, and could not be counted on to propel the ship. Forward motion was of critical importance because, during this storm, the *Globe* had to keep sailing up the sides of thirty-foot waves. If she was unable to do this, she'd lose her headway, slide back down, and be inundated by the following sea. In a fifty-mile-per-hour gale, the main topsail provided sufficient propulsion to keep her moving. The foresail, lowest and largest on the foremast, helped in a different way. As the ship drove along, this was one of the sails that actually exerted a lifting influence, preventing the vessel's bow from plowing into the waves and shipping a sea.

The *Globe* steered by a type of system called the *shin cracker,* in which the steering wheel was mounted directly on the tiller, the thick timber that projected forward from the top of the rudder, acting as a lever by which the rudder was moved. The steering wheel was simply the forward end of a horizontal capstan mounted on the tiller head. Ropes wound around the capstan to relieving tackles on either side of the ship, so that when the wheel turned, the rope pulled the tiller, and the whole assembly — rudder, tiller, and steering wheel — moved in an arc a few degrees one way or the other, right across the deck. It was called a shin cracker because when a big sea walloped the rudder, this motion would be transmitted to the tiller, causing the wheel and tiller head to lurch unexpectedly in its arc across the deck. The name, by all accounts, was richly deserved.

During a gale, and especially with a following or quartering sea, George Comstock, William Lay, or any of the other boys in the crew would not have had the strength or skill to manage the wheel alone. At such times, two men would be assigned this duty, probably with a mate standing by to advise and assist. It was vitally important to keep the ship headed in the direction of the waves, wind at her back. If she *broached to,* or got sideways to the seas, she'd fall into a trough, lose her propulsive power, and be swamped.

Even with Worth's "skillful management of the helm," plenty of water was sure to come aboard. There was no heat on the ship, and after a four-hour watch or a two-hour trick at the wheel, a man would be soaked to the bone. He'd crawl shivering into his bunk, and hope that his body heat would dry his clothes. The cook's shack, or *caboose,* lashed down in some out-of-the-way corner of the main deck, was continually drenched. Lighting a fire, assuming some dry kindling could be found, was sufficiently dangerous to be out of the question. Even if there had been a cook fire, all that rolling and pitching would have made hot liquids more of a hazard than a comfort. So, for two days at least, the men stayed shivering and wet, getting by on cold food and water.[1]

If a reef needed to be put in the topsail in order to reduce its area (hence its exposure to the wind), four or six of the crew would be bawled up from the forecastle and sent out on that yard on footropes, feet thrust behind them, chests against the yard, holding on for dear life while trying to pull the slackened bight of canvas over the top of the yard, to then tie the reef points around the furled portion of the shortened sail. It was not an easy operation in the best of conditions. On the *Globe* that New Year's Day, it would've been performed by a gang of teenaged boys, dancing on a rope in a howling gale.

By the end of those forty-eight hours, every one of them was well on his way to being a seaman.

Rather than beat down the coast of North America, it was customary for whaleships to follow the Gulf Stream and the prevailing westerlies across the Atlantic to the Azores, where they'd stop for fresh provisions. However, the *Globe* probably passed south of these islands during the gale. A week after the storm, her crew caught sight of the Cape Verde Islands, a hazy hump bearing southwest at a distance of twenty-five miles. The Cape Verdes, several hundred miles off the West African coast, were also a customary provisioning port, but the *Globe* did not stop there either. Nor did she linger on the whaling grounds farther down the Atlantic in the vicinity of Saint Helena. Captain Worth was certainly aware that by this point in the *Globe*'s prior trip, he and Captain Gardner had already gotten two whales stowed down, but Worth had something else in mind.

Elmo Hohman, in his seminal study of American whaling, speaks of whalers commonly spending "a long summer cruising along the 30th parallel of North Latitude between 145 and 165 degrees West Longitude."[2] This was a vast area of the central Pacific between Japan and Hawaii. On the *Globe*'s prior voyage, Captain Gardner had stayed in the 30th parallel between 145 East and 170 West Longitude from April to October. He'd killed thirty-eight whales and stowed almost 1,200 barrels of oil during that time. This was an area the *Globe* had pioneered, and even part of a season there might yield the beginnings of another good trip.

If Thomas Worth proceeded without delay through less productive waters, he could reach this area, or even the Japan Grounds, by early summer.[3] He would then be in company with the rest of the fleet. This would mean more gams, the possibility of letters or news from home, and more information from other captains about where whales were. It appears that he was headed for the western Pacific with all due speed.

This is not to say that the *Globe* left Nantucket late, or that Worth felt she had. Given the impossibility of predicting when a ship would come in, and the vagaries of refitting and finding a crew, it was impossible to predict when any vessel would, or should, depart on her whaling voyage. But of the thirty-four dates of departure that Starbuck lists for Nantucket vessels in 1822, twenty-two were between May and September. It is possible that Thomas Worth felt he was behind the fleet. His failure to put in at either of the two traditional watering places early in the voyage suggests what his intentions were. The *Globe* swung a wide arc around the top of the Atlantic and, with the northeast trades off the African continent at her back, homed in on Cape Horn.

As the weather improved, the men settled into the dailiness of life at sea. There was always something to do aboard ship, and whichever watch had the duty was sure to be busy doing it under the eagle eye of one of the mates. Repairs had to be made after the gale. Cargo that had shifted in the rough seas needed to be restowed. Torn sails would be patched and bent, spars scraped and slushed, rigging repaired — wormed, parceled, served,[4] and set up, then tarred for protection from the water. The whaleboats, if they had not been painted before, would be painted now, black or white (blue and green were also common), each with its own identifying markings at the bow or distinctive color along the sheer strake. The whalecraft would be sharpened, sheathed, and sharpened again. If all else failed, there was always oakum to be picked. This task consisted of unwinding strands of old wornout rope that could then be used for caulking. The chore was, like many other shipboard tasks, labor intensive but functional.

Every day just before noon, Captain Worth would appear on deck and, if the weather permitted, take his sights. The sun's altitude at the meridian would give him the vessel's latitude. In the evenings he would take the angle of the moon and whichever fixed star happened to be within its path. This would be compared with the information in the *Nautical Almanac* to give Greenwich time. The difference between Greenwich time and local time would produce a fair approximation of their longitude. When bad weather obscured the skies, the *log and line* were used to assist in navigation by the ancient process of *dead reckoning*.[5] A floating object, the *chip log,* would be thrown overboard, and the line, with knots at regular intervals, paid out after it. A small glass, something like an egg timer, measured the rate at which the line went out. (The *Globe* had one that measured intervals of fourteen seconds.[6]) This would give the rate of speed of the ship, and Captain Worth would translate this into real distance, combine with direction, and, after subtracting the effects of currents and the slippage of the ship in the wind (or leeway), plot from the last known point to arrive at a surprisingly accurate estimate of his position.

The second dog watch each day, from 6 to 8 P.M., was when the men relaxed. Old hands whittled, yarned, or worked on carving bone implements for shipboard use.[7] Greenhands tried their knots and splices, learned the ropes, and *boxed the compass* since "the young seaman, in preparation for taking his turn at the wheel, had to learn to gabble off the points of the compass, in quarter points, both forward and backward."[8]

Most of the business on a whaleship was done in silence so that commands from the officers or cries from the lookouts could be distinctly heard. But in the relaxation of the second dog watch, the men on deck could converse normally. If someone had a fiddle or a flute, this was when he'd play it. Sunday, of course, was the day of rest. The men slept. Those who could, read — Bibles, letters, old newspapers from home, thumbed and grimy, pages worn thin from use.

On January 17, 1823, they crossed the equator. In the merchant service this first time, *crossing the line* was a benchmark event in the life of a sailor, and the greenhands were customarily subjected to whatever merry tortures the old hands could devise and the captain would allow. Generally, there was a ceremony in which Neptune, enthroned with his queen (the possibilities for ribaldry began here), initiated the sailors who had never before sailed across the equator. These rites might include hazing; beating; and "shaving" with slush, tar, and a piece of iron hoop. However, these rituals do not seem to have been a strong part of whaling culture. Some writers speak of it, some do not.[9] Neither William Lay in his book, nor George Comstock in his manuscript account, say anything about a ceremony when the *Globe* crossed the line. William Comstock is similarly silent on the matter.

But it was still early in the voyage. Captain Worth was interested in making good time, and Samuel Comstock was biding his.

chapter thirteen

The First Whale

South Atlantic, 1823

 Twelve days south of the equator, on January 29, 1823, the *Globe* experienced the kind of delay that Captain Worth was only too happy to encounter. From the lookout on the main crosstrees came a cry,

"Blooows . . . !"

Picked up like an echo from the foremast, "Ah, blow! Blows. . . . She blooows . . . !"

The effect on deck was electric. Every man froze at his task, eyes directed skyward.

"Where away? How many?" Worth was looking aloft with the rest of them.

As the direction and size of the pod were being reported, the off-duty hands rushed out of the forecastle companionway, rolling up their cuffs and tightening their belts, peering over the rail in hopes of catching sight of their quarry.

A sharp-eyed lookout at the crosstrees could spot a whale's spout five or six miles away. The baleen whale, with its two blow-holes, sent up what the whalemen called a *forked spout,* almost perpendicular to the horizon. The sperm whale was distinctive in that its single spout shot forward at an angle of about forty-five degrees. So, before Captain Worth lowered a boat, he'd know whether he was chasing sperm whales. If the school was a great distance away, he'd try to sail within a mile or two before bringing the ship about and lowering the boats. Every attempt was made not to *galley,* or frighten, the whales.

The masthead lookouts were replaced by one of the shipkeepers, who'd monitor the ensuing action. Line tubs and water and lantern kegs were put in the boats while they were still on their davits. Boatsteerers and boatheaders clambered aboard and lowered each of the three boats, sliding down the ropes (*davit falls*) to take their assigned oars. Then the boats fanned out in the direction of the school, Worth and Fisher in the starboard boat, Beetle and

168

Smith in the larboard boat, and Lumbert and Comstock in the waist boat. Because the whaleboats were at water level, the distant whales might not be visible to them. They relied, therefore, on signals from the ship set according to instructions from the lookout. Pennants and sails would be hoisted or lowered singly or in combinations according to the *Globe*'s prearranged code.

Any given sperm whale would exhale, or *blow*, a certain number of times, then submerge, or *sound*, for as long as an hour. When it surfaced again it would blow exactly the same number of times before it sounded. If this cycle were disturbed, a whale might swim some distance away or submerge briefly, but it would not sound again until its breathing cycle had been completed. It often happened that a boat would approach a whale that had completed its breathing cycle, only to have it sound before they could harpoon it. When that happened, the boatheader would have to guess where it would surface next and try to get his boat there. After the whale reappeared, the boatmen knew it would have to stay surfaced until it had completed its breathing. Sometimes the hunt was a gradual approach over several breathing cycles. If the boatheader guessed the whale's direction incorrectly, or if the submerged whale was moving faster than the boat, it could make for a long row. The ship might diminish to a dot on the horizon, visible only from wave tops. It might disappear altogether. Darkness or fog might come on.

When the whale was ready to sound, it would roll down head first, exposing its broad tail fins, or *flukes,* for a moment before disappearing. Or it might simply settle, dropping beneath the surface without a trace of movement. When the whale sounded in sufficiently calm water, the men would see an oily patch that it had left on the surface. This was called the *glip.* Some whalemen believed there was some mysterious sensory connection between the whale and its glip, and that if they rowed over this oily patch, they'd galley the whale. A whale might also *lobtail,* hanging in the water head downward and violently slapping its flukes on the surface of the sea, producing a loud report. On some occasions it might propel itself vertically out of the water, until half its body length was in the air, before falling back with a tremendous splash. This was known as *breaching.*

Because the whale's eyes were on the sides of its enormous head, its front and rear vision was limited. Head and tail, therefore, were the customary approaches to the whale, with a frontal approach, *going head on head,* preferred because it avoided the flukes. Eventually the men would draw near enough that, when the whale spouted, the mist would rain down upon them, an acrid, foul-smelling condensation of the moist expended air from the creature's lungs. The oarsmen, whose backs had been to the whale all this time, would look over their shoulders uneasily as the boatsteerer, thigh braced against the *clumsy cleat* at the bow of the boat, made ready to dart his iron — six men on a splinter of cedar about to stab a creature whose size and power were on the order of a steam loco-motive. If the iron worked loose from the whale's flesh, whalemen said it *drew,* a common enough experience in the early days.[1]

But now the men in the *Globe*'s boat were fast to a maddened whale, and the first order of business was to get out of its way. The boatheader commanded "Stern all!" and then ordered the tub oar to "Wet line!" to prevent its charring as it ran around the loggerhead at the stern and

171

then out through the chock at the bow. The oarsmen would *peak* their oars, raising them out of the water, and the boatsteerer and boatheader would exchange places. In the pain and rage of having a harpoon stuck two feet into its back, the whale might *turn flukes* and sound, running line out of the tubs on its way to the bottom. Or it might run, at an initial burst of up to twenty-five miles per hour, leaving the men to hope the drag of the boat would exhaust the whale before their line ran out. Occasionally it happened that, instead of running or sounding, a whale would turn against its tormentors, slapping its flukes on the boat, upsetting it from beneath, ramming it, breaching next to it and crushing it, or chomping it to splinters.

This was deadly business. And the moments after harpooning and before killing, the two instances when the boat was in close proximity to the whale, were the most critical. It was here that the boatheader's or harpooner's ability with the steering oar could mean life or death to them all. On the *Globe*'s previous trip, a whale had "stove the waist boat some & knockd Mr. Oisten overboard" giving the mate a serious head injury. Two weeks later, a black man named Allen was killed, slapped out of the boat by the flukes of an enraged whale that had also "hurt the cook some."[2] There was also Ahab's fate to contend with. "A foul line, as it leaps like a thing of life from its coils in the tub, is . . . treacherous . . . if it catches the arm or leg of one of the crew, as it sometimes does, the unfortunate man would be carried out before any assistance could be rendered him."[3]

This time they were able to keep out of the whale's way, and the boatheader churned the twelve-foot lance deep into the whale's lungs, severing arteries and causing it to spout red. The whale would now be choking on its own blood. It would also, in its extremity, void the contents of its stomach and bowels. The messy death of such a gigantic mammal must have been a moving and terrifying spectacle. It was also, according to many accounts, nauseating. The arterial blood, fecal matter, and vomited squid combined to produce a sickening odor. However, this was the smell of success, and even the queasiest whalemen learned to tolerate and, eventually, to appreciate it.

The blood would come "thick as tar." The whale, in its *flurry*, would swim in an agonized circle and then, after thrashing the bloody water, roll heavily on its side or back, with its dorsal fin in the air. This was called *finning out,* and it signaled the end. If there was any doubt, the lance blade or one of the waif poles would be thrust into the creature's eyeball to be sure it was dead. Then, with the boat spade, a hole would be cut in the carcass, and the whale would be towed back to the ship.

The chase and the kill were without question the most exciting part of the whaleman's job. Dozens of witnesses have given us memorable accounts of the boats racing one another to their quarry; of the mates urging their crews on in spitting, hopping, roaring monologues calculated to turn them all into blood hunters; of the boat silently gliding up to the back of the lolling monster; of the mate's hissed command (inevitably, it was "Give it to him!") and of the harpooner's throw; of misses, rage, and dejection; of strikes and wild "Nantucket sleigh rides" if the whale ran, or creaking, bow-dipping minutes of tension if the whale sounded, running out fathom after fathom of line on its hell-bent way to the bottom; of the mate and his harpooner changing places in the fragile, tipping boat; of the potentially deadly approach to the harpooned whale; of the boatheader's thrust and the whale's spectacular flurry and death. But of the *Globe*'s first whale, William Lay says simply, "On the 29th . . . we saw sperm whales, lowered our boats, and succeeded in taking one." William Comstock makes no comment at all.

chapter fourteen

Cutting In

South Atlantic, 1823

It's a good bet Samuel Comstock didn't harpoon that whale. If he had, Lay would have mentioned it, or George would have noted it in his manuscript account and relayed the information to brother William, to whom he reported everything else Samuel told him. William, always eager to enhance his hero's career, would have included it in his biography. In addition, it was a big whale, a bull, probably lingering on the outskirts of the herd of cows and calves. Lay mentioned that it yielded seventy-five barrels of oil. Historian Obed Macy describes a typical sixty-barrel sperm whale as being sixty feet long, twenty-four feet in circumference, flukes seven feet wide, jaw fourteen feet long, five to nine inches of blubber all around.[1] This fellow could easily have been seventy-five feet in length. Bulls, when mature, tended to be loners. If they accompanied a herd, they stayed at its fringes.

The carcass, when they finally got it back to the *Globe*, was chained to the starboard side, flukes forward, and the ship positioned so that the whale would be to the weather. The sails would be trimmed, allowing the vessel to make slight headway, which would help keep the carcass close alongside. The force of the wind in the rigging would also make for a more stable working platform.

Then the toil began.

Standing on planks rigged fore and aft of the whale, or simply hanging over the side and using the beast's back as their working platform, Captain Worth and his officers wielded their long, razor-sharp cutting spades.[2] First, a semicircular incision about eighteen inches wide was made in the blubber immediately behind the head, just above the whale's fin. A hole was cut in the end of this strip, and a man was lowered over the side to guide a heavy metal *blubber hook* (or often a massive wooden toggle) through the hole. When this was secure, men at the windlass, in the bow of the ship, would begin heaving on handspikes or the brake of the windlass. The windlass brought in a heavy line that was rove through the *blubber tackle* — two large blocks secured to the mainmast just beneath the mainsail yard, and shackled to the blubber hook. As the tension on the line increased and as the men on the cutting stage hacked away at the blubber strip, the massive strip would come loose from the whale and be raised up the side of the ship toward the mainmast. In one of his brilliant descriptions, Herman Melville compared this *cutting-in* process to peeling an orange. As the crew at the windlass begins taking up tension on the line attached to the blubber strip,

> the entire ship careens over on her side . . . she trembles, quivers, and nods her frighted mast-heads to the sky. More and more she leans over to the whale, while every gasping heave of the windlass is answered by a helping heave from the billows; till at last, a swift, startling snap is heard; with a great swash, the ship rolls upwards and backwards from the whale, and the triumphant tackle rises into sight dragging after it

the disengaged semicircular end of the first strip of blubber. Now as the blubber envelopes the whale precisely as the rind does an orange, so is it stripped off from the body precisely as an orange is sometimes stripped by spiralizing it.[3]

All this was well enough in theory — the strip of blubber spiraling neatly down from head to flukes. In practice, whale and vessel were likely to be bobbing in the seas. The officers, standing on slippery boards six feet above the whale, were moving up and down as the ship rocked, and had to make accurate incisions on their target, which was also moving independently of the planks. One in five thrusts might be on the mark.

The man who went over the side to insert the blubber hook might also be secured by a *monkey rope*, which ran from a loop around his waist up to the deck, where it was held by a shipmate on whom his life depended.[4] In these latitudes, the bloody carcass would almost certainly have attracted sharks from miles around, and these ravenous creatures would make no differentiation between whale blubber and the man's leg or arm as he struggled to

179

secure the hook. A moment's inattention would result in a slack line; the man below might be washed off the carcass and into waiting jaws, or crushed between the whale and the ship. It was a dangerous and difficult job, and it often took many attempts before the hook was secure. The man on the other end of the monkey rope had to keep a constant tension on the line and be ready to give a lifesaving heave on an instant's notice.

When the peeled strip of blubber, or *blanket piece*, had been hauled all the way to the upper block at the mainmast, a second hook was fastened to it just above the whale's body, and another purchase was taken by a second blubber tackle rigged next to the first one. The top strip was then cut loose from the whale by a long, sword-like tool called the *boarding knife*, and swung aboard the ship, suspended from the main yard. It could weigh more than a ton, and as it boarded, it swept everything from its path.

While the second piece was being cut from the whale and hauled aloft, the first blanket piece was lowered through the main hatch into the blubber room on the lower deck, where the boatsteerers waited with their own cutting instruments. First they would remove all traces of flesh and muscle from the blubber to prevent discoloration of the oil. Then they'd chop the blubber into more manageable *horse pieces,* which would then be sent topside again to be sliced not quite all the way through into *books,* or *bibles,* or *bible leaves,* so named because the slices looked like pages in a thick book. These books, in turn, were thrown into the try pots to be rendered into oil.

The try works were nothing more than two brick fireboxes built over a protective reservoir of water. On top sat two 250-gallon iron pots partially filled with fresh water. A wood fire was lit in the fireboxes, the water was brought to a boil, and the blubber bibles were thrown into the pots. When the first pieces of blubber had been rendered, the refuse, or *cracklings,* would be skimmed off and thrown into the fire for fuel. Thus, as Melville observed, the whale consumed itself. The water would quickly evaporate and the pots would be boiling pure oil, to which more blubber would be added.

During this time, Cyrus Hussey, the cooper, would be stationed on the quarter-deck at his grindstone. Rowland Coffin or one of the other boys would be at his side, turning the grinding wheel as Hussey kept a razor edge on the dozens of knives and spades used in cutting up the whale. Cetacean blubber was more like steak gristle than lamb fat. It was tough and difficult to cut, and Hussey and Coffin kept to their wheel all through the cutting-in process. When the last cutting spade had been sharpened, these two could resume their primary task of preparing the casks to stow the whale oil.

As it boiled out, the hot oil was ladled into copper cooling tanks beside the try works, then into casks for further cooling. From there it was piped, via canvas hoses, into the large casks in the hold. The fires burned night and day until the last bit of blubber had been rendered. Eyewitnesses were unable to resist comparing the try works at night to hell, and indeed sweating, shirtless whalemen pitchforking hunks of blubber in front of darting red flames and thick, black smoke must have looked infernal.

The trying-out process was also, in common with so many other operations in the whaling industry, quite dangerous. Any fire aboard a ship had the potential for disaster; fire beneath 400 gallons of oil increased that potential tremendously.

The crew of the *Globe*, exhausted from killing the whale and cutting it in, had to exercise constant vigilance as long as the oil was being rendered.

While the bible leaves were cooking away, the mates, in a gargantuan feat of dissection, would sever the whale's head from its body and, after removing the jaw, stand the uppermost parts of the head — the *case* and the *junk* — upright, and either haul them aboard or secure them against the side in that position. This represented a quarter or a third of the whale's mass. From the case came pure spermaceti, the substance used in making candles. The junk was composed of meat and fat, but no blood. It would be boiled out and rendered into regular sperm oil.

The size of these creatures is difficult to envision; the heads of some of the whales from the *Globe*'s past trips yielded 300 gallons of spermaceti. If the whale was large enough, the most efficient way to extract its treasure was to lower a man into the case.

After cooling, the oil was ladled off into casks, which would be lashed along special rails on the weather deck until all the operations were complete, then properly stowed away below. If rough weather came up unexpectedly, the casks turned into so many potential missile hazards, rolling or flying across the decks.

Especially if it were a sickly looking or skinny whale (not the case with this bull), its intestines would be probed for ambergris. This substance was highly valued as an ingredient in the manufacture of perfume. It appeared in irregular lumps that were formed as a protective coating around a foreign body, such as a squid beak, that had lodged in the whale's intestines. Poking around the bowels of a freshly killed sixty-ton mammal doesn't sound appealing, but the whalemen had plenty of motivation. Ambergris was worth its weight in gold.[5]

After the last of the oil was ladled into the cooling tanks, the try-works fires were extinguished, and the ashes were used to make lye. On well-managed ships, this was combined with sand and copious amounts of elbow grease and seawater, and applied to the decks, bulwarks, hatches, and companionways, leaving them spotless. The only parts of the ship that wouldn't come clean were the sails. One or more courses or topsails might have been run up to steady the ship during trying out, and the dense smoke from burning cracklings would have stained them. Sooty sails were the mark of a whaleship.

It is at this moment in the voyage, during the stowing down of that first whale, that Samuel B. Comstock makes his initial appearance. According to William, contact with the whale oil caused Samuel great distress, "filling him with biles and inflaming his flesh."

One instinctively thinks of oil as soothing, and in the 19th century, sperm oil was considered to have medicinal properties. However, another whaleman-author says that constant exposure to sperm oil and salt water caused "painful tumors to break out over the whole body."[6]

In this instance, one of Samuel's grievances may have had a legitimate cause.

It probably took several days to try out those first seventy-five barrels and stow them down, but this was another whaling operation at which Captain Worth's men would improve. In 1820, under Captain Gardner, the *Globe*'s crew had taken three days to try out their first whale. By the time they got to Japan, they'd had as many as five whales tied alongside at one time and were stowing down one or more a day.

Even while the crew was cleaning and

stowing, Captain Worth made sail again. On February 23, the *Globe* passed the Falkland Islands. Now, as they approached the tip of South America, all hands were employed readying the ship for rounding Cape Horn, South America's southernmost extremity. Here the waters of two oceans met, and freezing Antarctic air was continually in collision with warmer, moister air from higher latitudes. It was an area of ferocious storms and high seas, and the difficulty of sailing its waters was legendary. Ships might spend weeks battling adverse winds and currents, clawing a few miles ahead only to be blown dozens back in the next gale. In earlier years, the crews, increasingly fatigued, starved, and frozen, would become less and less able to cope with the difficult conditions in which they were trapped. As a result, many a vessel met its end in these waters, overwhelmed by monster seas, grounded and beaten to death on a hidden shoal, or unable to clear the next fatal point of land.

By the *Globe*'s era, the routes and approaches, at least, were well known. Crews understood how to prepare, and the losses were fewer. Captain Worth ordered the royal yards sent down, the best sails bent, and the lighter sails, intended for tropical weather, stowed. The rigging was checked and reinforced, with preventershrouds got up to secure the masts. The boats were upended and lashed tightly to the davits. All cargo was stowed with extra care. Heavy clothing was broken out.

They seem to have made this troublesome passage in little more than a week. On March 1, they spoke the *Lyra*, a New Bedford whaleship that had left port just a day before the *Globe*. That they were six days south of the Falklands and still in calm enough seas to communicate with another vessel suggests they'd hit a patch of good weather. By March 5, they'd rounded the Horn and were headed north, no doubt feeling quite lucky to have gotten through with so little difficulty.[7]

After accomplishing this feat of navigation, American whaleships usually stopped at Valparaiso or one of the other provisioning ports on the western coast of South America. The men would get a chance to rest and thaw out. Fresh meat and vegetables could be purchased, and at the very least fresh water would be laid aboard. Then a captain might spend several months cruising in the vicinity of the Galapagos or the Offshore Grounds between Peru and Hawaii as the *Globe* had done on each of her last three trips.

When a whaleship is on the hunt, she moves at a leisurely pace, spending weeks working back and forth over a few degrees of latitude, thoroughly scanning every square mile for that telltale white puff of vapor. On her prior trip, the *Globe* had spent five months — November to April — between Cape Horn and Hawaii, and she'd stowed down ten whales in that interval. This time, Captain Worth bypassed Valparaiso and did not stay long on the whaling grounds west of Peru. Perhaps for this reason, the *Globe* saw only one whale and killed none.[8]

Thomas Worth wanted to get "on Japan" as quickly as possible.

chapter fifteen

The Golden Age

Japan Grounds, 1823

By May 1, 1823, after nearly five months at sea, Captain Worth and his crew reached the island of Hawaii, or Owyhee, as it was spelled then. During their approach to the island, the man at the masthead cried out a school of blackfish off the lee bow. Blackfish were a species of small whales, but they were a source of oil, and captains, particularly when nothing better appeared, would occasionally lower for them. Before making a decision, however, Worth resorted to his glass. The telescope (perhaps the same one he held in his portrait) revealed the prospective quarry to be canoes rather than whales.

Industrious islanders bringing produce

to trade manned these craft. That evening, as the *Globe* stood under easy sail ten miles offshore, they bartered potatoes, sugar cane, yams, coconuts, bananas, and fish for pieces of iron hoop and similar trade goods. On the following day, the *Globe* stood on and off, wearing slowly upwind and down, while more traders came aboard.

This somewhat contradicts the traditional image of languid, sensuous islanders, continually at their ease in a tropical paradise that answered all their material wants. The Sandwich Islanders had adapted quickly to American and European visitors, and had become aggressive traders.[1] Missionaries had recently arrived at the islands, and this entrepreneurial drive was equated with the introduction of Christian morality, all under the rubric of "civilization."[2]

Although the Spaniards may have visited these islands as early as the 16th century, the islands' first modern European visitor was Captain Cook, who arrived in 1778 on his third voyage to the Pacific. The dramatic events that accompanied Cook's voyage were recent enough to still be on the minds of the *Globe*'s crew. George Comstock, in his manuscript account, said of their anchorage that night, "We were off Carakakoon [Kealakekua Bay] just where Capt. Cook was killed."

Only forty-five years had passed since Cook's murder, but much had changed. The islands had been unified under the rule of the dynamic Kamehameha I[3] who, in 1789, made two western traders his advisors. Under this regime, Kamehameha's power and wealth increased, and the Sandwich Islands became a mercantile center. On his father's death in 1819, the selfish and ineffectual Kamehameha II abolished taboo and idolatry, an act which destabilized indigenous culture and opened the way for missionary activity. This coincided with the first visits from American whalemen, who vastly expanded commercial opportunities for the islanders and further altered traditional society. While the Sandwich Islands were a beehive of missionary activity, they also supported the amorality and lawlessness common to seaports the world over.

Along with provisions, those native canoes brought women.

The *Globe* did not land at Hawaii, and no crewmen went ashore. This was a tactic commonly employed by whaling masters to avoid the delays caused by rum, desertion, and troubles with natives. Certainly, it was consistent with Captain Worth's impatience to get "on Japan." However, these men had been at sea for five months, and the favors of the women of the Pacific islands were at this time one of the mariner's expected prerogatives.[4] William Comstock says that Captain Worth allowed the women aboard, though he explicitly forbade any of them to remain on the ship overnight.

Samuel Comstock ignored those orders. The next morning, William wrote, "Lady Comstock made her appearance, emerging from steerage, with an air of great dignity, dressed in a new Scotch bonnet, and rose blankets, which her gallant had presented her. The Captain stared at this unexpected apparition; but perceiving under whose protection she was, made no remark."

According to both William Lay and William Comstock, Captain Worth always favored Samuel. However, by allowing Samuel openly to flout his authority, Worth weakened his credibility and risked disrupting his relations with Beetle, Lumbert, and Fisher. If his officers couldn't rely on their captain for the consistent application of his rules, how could they be certain he'd back them up in a crisis? At the very least, such favoritism would have alienated Samuel, a recipient of special privileges, from the mates.

Something like this seems to have happened, at least between Beetle and Comstock, and doubtless it was exacerbated by other unrecorded exchanges. Gilbert Smith, the second boatsteerer, stated, "I heard Comstock say that if he ever lived to get to America he would be the death of William Beetle the mate. . . ." And George Comstock testified that "the first mate whipped me once with a rope's end."[5] This may have fueled Samuel's dislike for the first mate, or it may have been occasioned by Beetle's dislike for Samuel. George does not elaborate.

The *Globe* traded at Hawaii for two days, and sailed northwest to Oahu, known then as Woahoo. Again, the ship did not put into port, and no liberty was granted. They spent a day in Honolulu Harbor trading for more provisions, and stood off for the coast of Japan in company with the *Palladium* of Boston and *Pocahontas* of Falmouth. Two days west of the Sandwich Islands, the three ships shared a sperm whale and then went their separate ways.

Worth and his crew reached the whaling

ground by early June, and spent the summer and fall of 1823 "on Japan." They'd arrived at the proper season and gammed with several American whaleships, including the *Enterprise* of Nantucket and their old friend the *Lyra* of New Bedford. But, despite Worth's best hopes, the *Globe*'s four months there were not a complete success. After almost a year at sea, they'd stowed down only 600 barrels of sperm oil — about half of what had been done by that point on the ship's previous voyage.

During these months of hard work and unspectacular results, the situation aboard the *Globe* began to deteriorate. Samuel Comstock did his best to rock the boat.

There was, for example, an incident on the Japan Grounds. During the gam with the *Enterprise*, the whalemen, as was the custom, entertained themselves. In the spirit of the moment, Comstock challenged the third mate, Nathaniel Fisher, to a wrestling match. Fisher, at five feet, eight inches, had a two-inch height advantage, but the boatsteerer was broad-shouldered and muscular. To Comstock's considerable surprise, Fisher put him down easily. That should have been the end of it, but, rather than accept his defeat good-naturedly,

Samuel became enraged. William Lay says Comstock struck Fisher, whereupon Fisher "laid him upon deck several times in a pretty rough manner." Cooler heads quickly prevailed. Mate Fisher forgot the incident immediately. Samuel did not.

Stephen Kidder reported that "Constant Lewis was one put in Irons for differing with Comstock the Boatsteerer and let out next Day — and the Captain . . . struck the cook [John Cleveland] another time on account of Comstock also."

While the *Globe* was on Japan, several men reported problems with the rations. Some of the meat may have been bad, and the manner in which it was served was very irregular. William Lay says the trouble stemmed from the size of the portions allotted to the crew, "the quantity sometimes being more than sufficient for the number of men, and at others not enough to supply the ship's company." Captain Worth again beat Cleveland the cook, this time for being drunk; his drunkenness may have played a part in the food problem. George Comstock says that not only were the rations irregular, there was never enough time to eat them. "When we went down to meals we would scarcely have the victuals brought down before the 2nd mate

[John Lumbert] would come down with this cry, 'Where are you there forward. Come out of that or I will be among you!' and if we did not start immediately the mate or Capt. would come along also and with oaths threaten of breaking our backs." Lumbert, nearly six feet tall, was the biggest man on the ship. He and Fisher were probably the disciplinarians.

William represents these difficulties as playing into his brother's hands. "When any man complained to Captain Worth that he was suffering from hunger, he would tell him to eat iron hoops. . . . It does not appear that our hero discountenanced these arbitrary proceedings; but, on the contrary, did all he could to encourage them; anticipating . . . effects favorable to his schemes."

As the voyage dragged to the wearisome end of its first year, and the *Globe* returned from Japan to the Sandwich Islands to reprovision, all the ingredients for insurrection were there. The crew had experienced indifferent success, bad food, capricious exercise of authority by an inexperienced captain, bullying and physical beatings from the officers, long confinement aboard ship with no liberty, and the concerted, pernicious influence of a malcontent. To these trials were added the inescapable factors of extreme occupational hazards, squalid, unsanitary living quarters, excessive tedium, long hours, and low pay.

The surprising thing is that these conditions were not unusual in a whaling voyage. In fact, they were the norm.

In this golden age of whaling, men could anticipate exactly the sort of treatment the crew of the *Globe* was receiving. All of the survivors of this voyage, after reporting the bullying, the beatings, and the difficulties with the food, stated that Captain Worth had treated them as well as could be expected.

Shipboard discipline was commonly enforced by tongue lashing, punching, kicking, or beating with whatever was handy, usually a rope's end or stick of wood. These were almost always unpremeditated responses to poor performance or minor breaches of discipline. A doltish or laggard seaman might also be subjected to the public humiliation of being sent up into the rigging for a certain period of time. This corresponded to being told to "go sit in the corner," and it is arguable that most shipboard punishment was commensurate with the level of discipline that might occur in the 19th-century home. As well as a prison warden and shop-boss, the whaling master was a father figure to the young men on his ship. He was called the "old man," and he usually meted out discipline the way their fathers had. Of course, the mitigating aspect of the mother's presence was usually absent, and in this case Thomas Worth was a first-time father.

A whaleship was an enclosed society, and when an offense reached a level that threatened the social fabric, sterner measures would be taken. Next in order of severity was flogging, a formalized, public punishment in which a man was tied to the rigging and whipped. The man administering the flogging used an unraveled rope or, more painfully, a *cat*, which consisted of nine knotted cords secured to a handle. The punishment was delivered in doses between one lash and multiple dozens, with the shirt on or removed. Alternatives to flogging included confinement in some small and enclosed space in the ship, without rations. Being *seized in the rigging* was the term for tying a man onto the ship's rigging with light *seizing* line. It combined public humiliation with the pain of being rendered immobile in an uncomfortable position. Irons were the most common alternative to flogging and, according to whaling historian Britton Cooper Busch, were quite effective, in the short term at least, in getting unruly seamen to modify their behavior.[6]

It occasionally happened that men found themselves aboard a hell ship, lorded over by a particularly sadistic master or mate. But even with a humane captain, the simple fact was that shipboard life was brutal and unpleasant. The only possible responses were submission, escape, or rebellion.

Submission to the rigors and injustices of life aboard a whaleship was the choice of the majority, but it was not a large majority. Many crewmen opted to jump ship. "Only in the rarest instances did a vessel return from a voyage with substantially the same crew with which she sailed. It was not unusual for a whaleship which never carried more than thirty-five men at any one time to come into port with several times that number of names in her account-books."[7] Elmo Hohman, Busch's predecessor in the study of labor conditions in the whaling industry, states that the average whaler might lose two-thirds of her crew to desertion over an entire voyage.[8]

So common was this practice, that it was regarded as a rite of passage. Clifford Ashley suggests that officers were likely to have some empathy for recaptured deserters because the chances were that they had in the past done the same thing. Both Ashley and Busch sum up the situation by referring to the same quotation. "If a ship were bound for Heaven and should stop at Hell for wood and water some of the crew would run away."[9]

On November 7, 1823, after her season "on Japan," the *Globe* returned to Oahu and six of her crew deserted.[10] Perhaps they left the ship because of ill treatment, bad food, or mere whim. Perhaps they had some inkling that their ship was bound not for Heaven but for Hell.

chapter sixteen

Abandoned Wretches and Cruel Beings

Honolulu, 1823

 Captain Cook by-passed Honolulu when he discovered the Sandwich Islands, but as early as the 1790s, British mer-chantmen were taking advantage of the capacious natural port there. Honolulu Harbor offered the best shelter of any in Hawaii, with sufficient room for hundreds of ships to anchor. It soon became a depot and resting place for Boston-based China traders, carrying sea-otter skins from the Pacific Northwest. When this branch of trade declined, those same American fur traders began exporting Hawaiian sandal-wood to the Canton market. After the death

of Kamehameha I, this valuable resource ceased to be controlled, and exports sky-rocketed.

By 1825, sandalwood was virtually extinct in the Sandwich Islands. Greedy chiefs and exploitative traders[1] had sent every stick of it to Canton, where the market glutted and prices crashed. However, a new source of commercial activity had already appeared on the scene. In 1819, the first American whaleship entered Honolulu Harbor. By 1822, more than sixty anchored there on their voyages to and from Pacific whaling grounds. During these years, American missionaries proliferated. Hiram Bingham and his Congregational associates made Honolulu their base, and set about attempting to offset the demoralization introduced by Boston traders and Nantucket whalers. All these émigrés brought American goods and American ways with them. Reverend Bingham even had a foursquare New England frame house shipped in pieces around the Horn. Honolulu soon became a bustling, polyglot port where the South Pacific and New England met — amicably, more often than not.

Into this lively situation sailed the *Globe*, exhausted from her year at sea. The broad, flat plain that surrounded the harbor would have been lined with shipyards and outfitters eager to provide for the ship's needs, and with fleshpots and gin mills to provide for the needs of her crew. In the background, church spires testified that Bingham and his friends were already busy winning converts and "civilizing" the islands.

Six of the *Globe*'s crew thought this interesting mix would be better than another two years of Worth, Comstock, and company. Daniel Cook, the mulatto and possibly the ship's steward; Holden Henman of Canton, Massachusetts; Jeremiah Ingham, Lay's friend from Saybrook, Connecticut; Paul Jarrett, also listed as a mulatto;[2] Constant Lewis, the last-minute addition to the original crew; and Joseph Prass, the fourteen-year-old cabin boy, all jumped ship at Honolulu. John Cleveland, the cook, was discharged.[3] Presumably, all the men who deserted were members of the same watch and hence on liberty together. Perhaps some of the men were part of Comstock's waist-boat crew, harried by him into deserting.[4] Two of them were recaptured in short order and returned to the ship in irons. But, says George Comstock, evoking the image of a certain wily, pint-sized Portuguese cabin boy, "One of the men having very slender hands slipped his irons and let the other out."

Outright revolt, or mutiny, was the course least often chosen by unhappy crewmen. The penalties for mutiny were severe, hanging being the most extreme. And, if successful, the mutineer was a marked man who could count on being hunted for the rest of his life or on removing himself from civilization altogether. If this was not sufficiently discouraging, masters and mates had access to firearms, and there are several instances of shipboard insurrections being put down by gunfire.[5]

Furthermore, there was one aspect of mutiny that ran against every tenet of a seaman's existence. Shipboard life was necessarily highly ordered, regulated, and ritualized. It was dominated by the presence of a clearly delineated chain of command, which, officially and unofficially, ran from the greenest foremast hand up to the captain himself. Everyone knew what his job was, and everyone knew where he stood within the hierarchy. Gilbert Smith, for example, would know that he was not expected to sweep the decks each morning, because he was a boatsteerer. A seasoned hand like Peter Kidder might get the best part of that duty, or avoid it completely, while the life of a greenhand like William Lay would consist of one menial chore after another. Although the highly rigid order of things may have been a surface irritant, these ritual observances served the comforting function of locating a man in the little universe that was his ship. The retired sailor who can't get nautical ways out of his blood is a comic literary convention, but his type was real enough. There was a deeply satisfying aspect to the regularity of shipboard life.

Nautical custom did not spring into being arbitrarily. The reason that the chain of command so precisely located every person was mortally important. In emergencies, every man had to know what to do. He'd know this by being told by someone higher up the chain. Frequently, the safety of the entire ship depended not only on the order coming down in a clear and timely manner, but also on its being executed immediately and without question. Thus, when Captain Worth ordered a change in course, the helmsman was required to repeat the order aloud, precisely as given, to ensure that he'd heard it correctly and would execute it exactly. When he was relieved, he would pass along his course to the man who relieved him, and this man in turn would repeat it aloud. Failure to do so could compromise the ship's safety. It also might result in a tongue lashing, a box about the ears, or even a rope's end — unpleasant memory aids, but with a life-and-death reason behind them.

In the years between 1800 and 1850, Starbuck lists over 7,000 whaleship departures. For all these voyages he records sixteen mutinies, a mere two of which occurred prior to 1825.[6] In most of these incidents the crew simply abandoned or destroyed the ship; few resulted in deaths; and only one in mass murder.

What would the deserters have done ashore? The possibilities were few.[7] Even if little Joseph Prass and the rest of the Globe's deserters were successful in evading Worth or the Honolulu authorities, the pleasures of sailortown would probably have paled quickly. When a fugitive's scanty funds were exhausted, he could expect small welcome in the barroom or the brothel. Of those who tried living among the natives in some more removed spot "on the beach," few were able to last long. Most Yankee seamen didn't care for the monotonous diet and climate, alien language and culture, and simple isolation. Eventually, they'd ship out on another vessel.

Captain Worth chose his replacements from just such a pool of men. In a sense, most of them were outlaws, sailors who'd illegally broken the terms of their shipping articles and left the vessel to which this contract bound them. Some may even have deserted several times. There was a sub-class of deserters, the *beachcombers,* who would ship aboard a vessel and stay with her until she reached the next interesting port, then desert again.[8]

The *Globe*'s particular group of replace-ments seems so spectacularly ill chosen that one has to wonder about Thomas Worth's grasp of human nature. George Comstock says, "The crew who had shipped at Woahoo were a rough set of cruel beings." William Lay calls them "abandoned wretches." Surely, on all the island of Oahu, in all the teeming dives that lined the harbor's dusty streets, there had to be more men than just these seven. If Worth picked them himself, he did a terrible job.[9]

The replacement crew seems so well suited to Samuel Comstock that the question arises of his involvement in their selection. Could he have driven the men of his watch to leave the ship, and then caused them to be replaced with men of his own choosing? Comstock had the respect of his captain, and at times he seemed able to bend Thomas Worth to his will. Of the seven men chosen, five were involved, at least peripherally, in the mutiny. Samuel had known one of them, John Oliver, from his adventures years before in Valparaiso. However, although William says that Samuel assisted the deserters in their escape, he does not mention his brother having any input in the choice of replacements. Rather, he reports that Samuel himself requested a discharge from the *Globe*.

This was a ruse. Samuel must have known that Captain Worth would never grant such a request. Discharges were far less common than desertions. A man with an illness or lingering incapacity, such as Cleveland's drunkenness, might be sent ashore, but given the expense involved when discharging a seaman in a foreign port, owners would choose to keep a man on board.[10] Occasionally, captains might swap unhappy crewmen the way sports teams trade players, but parity would be critical in any such trade.[11] The chance of finding two unhappy men of identical abilities was slim.

None of this applied to Comstock. In his capacity as a boatsteerer, he was one of the *Globe*'s best men. Worth had just lost seven, and was in the process of shipping aboard seven more of questionable value. He wasn't about to lose another one. William provides his brother's reaction to Worth's refusal. "He was much incensed, and was heard threatening, (in Jack Crown's shantee) that if Captain Worth took him to sea again in the *Globe*, it would be at peril of his life."

This was a warning as much as a boast.

So, from places like Jack Crown's shantee came Silas Payne of Sag Harbor; John Oliver (Samuel's running mate in Valparaiso) from England; Thomas Lilliston of Virginia; William Humphries, a black man from Philadelphia; Joseph Thomas of Norwich, Connecticut, "Joe Brown," a Kanaka, or native Hawaiian; and Anthony Hansen, an Indian from Cape Cod. Of these seven "cruel beings" only Hansen and the Hawaiian had no knowledge of, or involvement in, the mutiny on the whaleship *Globe*.

chapter seventeen

Comstock's Creatures

Pacific Ocean, 1824

The five of them were trouble from the start. William described Payne as "a tall, stern and reserved person, of some resolution, and a rebellious nature." John Oliver was a "little contemptuous fellow, coarse, vulgar and ignorant." Being new, and rough, they kept to themselves, aloof from the youngsters in the forecastle. Some of them probably filled out Comstock's whaleboat crew and served on his boat crew watch. George says they were "continually murmuring about the food, usage, etc." Peter Kidder would later testify that those men had a plan in place for a month before acting on it. Within a few weeks of their coming aboard, Payne, Oliver, Humphries, Lilliston, and Thomas were seen to be Comstock's creatures.

On December 9, they departed Honolulu for a "season on the line."[1] Captain Worth headed south, to a long, narrow band of uniformly warm water that followed the equator for several thousand miles through the mid-Pacific. Famed oceanographer Matthew Fontaine Maury, in his *Explanations and Sailing Directions*, presents a whaling captain's theory about why sperm whales congregated there: "It was to meet the bait brought down with the current, particularly near the equator in the Pacific, where a current is always found setting to the westward, and the whales generally found stemming it to the eastward."[2] The "Winds and Routes" chart in Maury's book shows how the northeast trade winds prevailed above the equator and the southeast trades below it — gentle, insistent winds pushing surface waters westward into the equatorial stream, against which the sperm whales swam, always alert for the squid that formed their diet. For the past five years, Nantucket whalers had been methodically working this ground.

George Comstock says, "We run down to the southward of the equinoctial, in about 2 South latitude we cruised a short time for whales, then shaped our course northward being bound for Fanny's Island." This malapropism referred to Fanning's Island, situated a few degrees north of the equator, and about 2,000 miles due south of Hawaii. It had been discovered in 1798 by an American pioneer South Seas trader named Edmund Fanning. Captain Worth might have planned to refresh the water casks there, or just to cruise in that vicinity, but as he made for the island on January 25, 1824, events overtook him.

Joseph Thomas had been named steward when he came aboard, and his duties included provisioning, catering, and keeping things orderly in the cabin and in steerage. Thomas, however, proved no better at this job than the man he replaced. The black man, William Humphries, who may already have been serving as cook, took over the steward's job, and Thomas was sent to the forecastle as a regular seaman. George reports that, a short while after Thomas's transfer, Captain Worth ordered the men on deck to tighten the fore brace. "He spoke to one of the men somewhat in this manner — I will knock you to h—l if you don't come out of the forecastle in a little more haste the next time. The man told him he would pay for it if he did."

This unnamed man was Joseph Thomas.

It was not unusual for a captain to change his personality once a whaling voyage began. A mild-mannered family man ashore could become a tyrant at sea, and this seems to have been the case with Captain Thomas Worth. That dandyish, capable-looking man portrayed in the painting at Martha's Vineyard was now edgy and volatile. In his defense, this was his first voyage as a master, and there were many pressures on him — the expectations of the extended Worth clan back on Martha's Vineyard; of his father, in whose house his wife now lived; and of the Mitchell partners who'd given him this opportunity. Perhaps, too, some of the pressure was self-generated. If he was driving his men, he was driving himself just as hard.[3] The result was a short fuse and a tendency to relate to his crew in a violent manner.

As William Comstock observed, Lay and the other teenagers in the forecastle accepted this behavior. For the greenhands, it was the only kind of treatment they knew aboard ship. They would have been "easily awed" by the officers, and they were still too attached to their homes and families to think of deserting. Joseph Thomas and the other grizzled replacements would have had a different reaction to the captain's outbursts.

"I told him it would be a dear blow for him," Thomas later testified. "He struck me once with a rope. I ran from the Chief, and the 2nd mate ordered me back, & the Capn took me by the throat & struck me 13 or 14 times with the end of the main buntline."[4]

The initial scuffle with the captain had attracted the attention of everyone on deck. They assembled aft, appalled by the beating but not daring to raise an objection. Says George, "One of the boat-steerers came and told us to revenge it and he would see us out, but we did not want to do anything to produce a quarrel with the officers."

There was Samuel Comstock.

Thomas was sent down to the forecastle, his back bruised "in a shocking manner." Down below, the wounded man among them, Silas Payne continued the rabble-rousing, telling the boys they must not be afraid to speak out. If the captain flogged any more in that way, they must go up and tell him to stop.

Samuel next began the process of sounding out the crew. Later that day, on lookout at the masthead with William Lay, he said, "Well, William, there is bad usage in the ship — what had we better do, run away or take the ship?"[5] Young Lay made no response. He was terrified at the suggestion and tried to seek out the second mate, but Comstock or one of his friends always seems to have been in the way.

The *Globe* was ninety-five feet on deck. Her crew would have had the run of the forward half of this space, an area about forty feet long and twenty wide. If two or more crewmen wanted to have a private conversation, they might resort to the hold below or go up into the rigging, where they'd be out of hearing and concealed by the billowing sails. But if one of the hands hailed an officer and requested a word, it would normally have to be done on deck, and it would be a highly visible action. William Lay, knowing he would have to answer for it in the isolation of the forecastle to someone like Payne or Oliver, would not have wanted to take the risk.

On the second dog watch of the evening after Thomas's beating, the men were lounging in the forward part of the ship.

215

When Beetle ordered them down to the fo'c'sle to take their supper, Comstock passed close to Thomas and asked if he would join them in the cabin that night. Several of the crew witnessed this brief interview. Joseph Thomas himself reports it, but not even he gives it any more specificity than that. Clearly, Samuel Comstock had approached Thomas on this matter before. Just as clearly, Comstock and his cohorts had some plan in place to confront the captain. Here, he was letting Thomas know that the hour had arrived. Thomas later claimed he didn't know that Comstock had murder in mind, and this is credible. The threat of a concerted work stoppage or the tactic of presenting the captain with a list of demands was a fairly common response to bad conditions. In any event, Thomas deflected Comstock's invitation, saying he "would think of it another time."

In these rapidly deteriorating circumstances, desertion might have been an attractive option. George Comstock, whose fraternal loyalty did not extend to whatever mayhem Samuel had in mind, says that he and "several" shipmates were planning to steal a boat and desert on Fanny's Island. "I had been very well used by the Capt.

and had nothing to complain of but fore-seeing there would be some noise yet between the crew and officers I was deter-mined to leave the ship." George's escape plans probably included seventeen-year-old Rowland Jones. Of all the hands who stood at the waist witnessing Joseph Thomas's beating, Jones was the only one who spoke up, asking the crew how they could stand to see a man being treated so.

On a chip of wood floating in the endless Pacific, one malcontent drives men to desert, even assists them in doing so. He knows one of their replacements. Others come to know him while on shore, and are drawn to him. They are brought together by their hatred of the captain. This captain is short-tempered and rough with the men and has been resorting more and more to physical violence. The malcontent and his core group spread the seeds of resentment and lead the crew in complaints about the food, the lack of time to eat it, and the bru-tality of the captain and officers. He and his henchmen develop a plan, perhaps beginning as a confrontation and escalat-ing to mutiny as their confidence grows.

The officers, meanwhile, have no idea anything unusual is going on. The replacements are a hard lot, and the crew is unhappy. But this doesn't matter to them. They are confident they can keep the crew functional and will use whatever measures it takes to do this. Their main business is killing whales and stowing down oil. Once they raise their next whale, the other problems will not be important.

The trouble is, no whales have appeared. The malcontent cannot be distracted from his grievances. His cronies are fugitives, the dregs of humanity. When the captain administers a severe beating to one of the crew, Samuel B. Comstock senses that his moment has arrived.

chapter eighteen

Consul Hogan

Valparaiso, 1824

 Among the consular records stored at the National Archives in Washington, D.C., is a letter from Michael Hogan, the American consul at Valparaiso, to John Adams, America's secretary of state. It was sent in August 1824, and it views the troubles aboard the *Globe* from a different perspective.

Sir:

> *It is a very painful part of my duty to lay before you the testimony of . . . part of the Crew of the Whaling Ship* Globe, *Thomas Worth the Master . . . which was rose upon on the night of the 26th of January last . . .*
> *It does not appear that there was cause*

to complain of the Captain of this Ship, which is uncommon, for in Justice to truth, my experience for upwards of three years in this Port, has proved to my conviction that the Masters of Merchant Ships trading here are oftener in error than the Sailors, who by severe inconsiderate and unfeeling treatment are driven to insubordination and desertion, then by a residence in the Heart of all sorts of Vice become destitute.

Masters promise much at home, in order to get a Crew, but never perform abroad. If a Consul insists on anything in favour of a Sailor or in obedience to the Laws laid down on the Articles, he is considered an enemy to the Captain . . .

I speak feelingly and with truth, but no fear of their frowns nor hope of their favours shall ever induce me to change from the course of supporting to the best of my power and abilities the measures best calculated to insure an obedience to our Laws and protect the Credit of our Country . . .

*With perfect respect
I have the Honor to be
Sir Your obedient &
very faithful Servant
Michael Hogan*

chapter nineteen

The Loose Hatchet

January 26, 1824

 At a critical moment, the officers were distracted from the trouble brewing around them. Their ship had once again fallen in with the *Lyra* of New Bedford. The two vessels had first spoken one another off Cape Horn the year before, and had fished together on Japan. Now, *Lyra*'s master, a man named Reuben Joy, came aboard the *Globe* for a gam with Captain Worth. If things followed true to form, Worth's officers — Beetle, Lumbert, and Fisher — would reciprocate by rowing over to the *Lyra* for a gam with Captain Joy's officers.

The two skippers must have discussed the luck they'd had so far, and made their plans for the coming campaign. They

221

agreed to sail together for a time. If the *Globe* tacked during the night, the watch would set out a lantern as a signal for the *Lyra*, so that she might follow. Captain Joy stayed aboard most of the day, but by the second dog watch, the gam had ended, and First Mate Beetle was back on the *Globe* to call the men to their evening meal.

Boatsteerer Smith's larboard-boat watch took the deck at 7:00 P.M. and stood until 10:00 that night. The captain came up once, about 8:00, to ensure that all was well. He had two reefs taken in each topsail, left orders to keep the ship *by the wind* (sailing as closely as possible to the direction from which the wind was blowing), and then went below. At 10:00 P.M., Smith's watch was relieved by Comstock's waist-boat watch, which would stand until 2:00 A.M. after which they'd be relieved by Third Mate Fisher and his crew.

On Samuel's watch, his younger brother George had the first turn at the wheel. This duty at the helm was sufficiently demanding to be broken into two *tricks* of two hours each. George would therefore be expected to stay at his post until midnight. As he came on duty, he was ordered by his brother "to keep the ship a good full." This was a repetition of the captain's orders. It meant she was to sail as close to the wind as she could, while still keeping her sails full. If, for example, the *Globe* were under the influence of a steady north wind, her course would have been "East by North" or "West by North." If the helmsman headed her any closer to the wind, the air would spill out of her sails, and the windward edge, or *weather leech,* of the sail would begin to vibrate, or *shiver.* The *Globe* was going along under double-reefed topsails that night, suggesting it was blowing quite a bit. George would have made sure he kept her a good full by paying attention to the main topsail. If it started to shiver, he'd bring the ship off the wind until the sails filled again.

This was the kind of shipboard evolution that even a greenhand like George would be able to accomplish after a year at sea. Still, Samuel came aft several times that night to check up on him. "He came to me several times and cursed me because I had the ship in the wind, but I took no notice of this." George could see plainly enough whether or not the ship was on the proper course. Samuel was in fact moving about the deck, making preparations for what was to come. The corrections in course were a pretense, and his curses betrayed his nervous excitement.[1]

At midnight, when George's turn at the wheel was done, he picked up the *rattle*. This was a noisemaker used by the helmsman to summon his relief, since he could not leave the wheel himself. He was just beginning to shake it when his brother appeared out of the darkness and uttered the statement that Melville deemed sufficiently terrifying to quote in *Moby-Dick*.

"If you make the least damn bit of noise I will send you to hell."

These words had the desired effect. Though George could not at that moment imagine the horrible specifics behind them, he grasped the fact that they carried the chill of death. "I . . . was suddenly checked by a brother in flesh but not in heart . . . little did he think I would ever get home to tell the fatal news."

He hung on to the wheel for dear life. Two feet in front of him, the lamp illuminating the binnacle compass provided a small, dim sphere of light. Fifty feet over his head, the main topsail might just be discernible as it obscured the night sky. The sibilant rush of the ship's hull through the water was the only sound to be heard. Otherwise, silence and blackness were all he had for comfort.

Samuel reappeared, now sufficiently pre-occupied as to be unconscious of the boy at the wheel. He lit a lamp at the binnacle and carried it down the booby hatch into steerage, where he would gather his weapons and rouse steward Humphries. Realizing something terrible was about to happen, George picked up the rattle a second time, in hopes of waking someone who could offer aid or support. But before he could get a sound off, his brother came up again, carrying something that he placed on a workbench near the cabin gangway. It was a boarding knife, the razor-sharp, sword-like tool used to cut through the wide blanket of blubber as it came off the whale and swung aboard ship. The blade of the boarding knife was about four feet long, pointed, with both edges sharp-ened, mounted on a wooden handle several more feet in length. It was a dangerous tool.[2] Samuel had appropriated several of them, and was laying this one by, in case of a later emergency. Then, says George, Samuel took his lantern and "went in the waist" of the ship — probably on the weather deck forward of the mainmast.[3]

From his position at the wheel, George's view would have been obscured by the cabin companionway, a five-foot-tall, closet-like structure built on deck to protect the entrance to the captain's cabin on the deck below. Also, the mizzen- and mainmasts, their attendant lines, and miscellaneous deck gear intervened. Samuel's lamp would have shone briefly off the rigging, and then all would have returned to blackness as he went forward.

Here Samuel met Silas Payne, lanky and taciturn; the stumpy John Oliver; the black steward William Humphries; and the Virginian, Thomas Lilliston. One or more of these men might have been in Comstock's watch, and so could have moved about without arousing suspicion. But how much suspicion was there to be aroused? It was after midnight. The rest of the men had turned in hours before and would, in all probability, be sleeping soundly. If Oliver or Lilliston had crept out of the forecastle, it wouldn't have been noticed. Likewise, Gilbert Smith, Rowland Coffin, and Cyrus Hussey would be asleep in steerage when Comstock and Humphries left the compartment. All the mutineers had to do was move quietly.

Comstock had armed himself with a tool described as an "axe." This could have been a cooper's axe, used for trimming barrel staves, or it could have been one of the hatchets from the whaleboats, which were used, among other things, to cut the whale line in an emergency. Such a tool would have been swung with one hand, rather than with the two hands it took to heft a woodcutter's axe. Comstock's "axe" was the primary murder weapon, and he swung it with one hand.

The five conspirators crept forward to the companionway that led down to Captain Worth's cabin. At this point, Thomas Lilliston abandoned his mates and, in George's words, "ran forward and turned in, not willing to have anything to do with killing the officers . . . he agreed to go, but finding them bent on the work he retreated."

They were indeed "bent on the work."

George, mute from terror, must have seen the four armed men go down into the cabin.[4] He was standing five feet from the cabin companionway and directly above the scene of the crime. Just in front of him, the binnacle compass was sheltered in a small structure that was open to the room below, so that the captain, simply by looking up, could ascertain the ship's bearing without leaving his cabin. If Robert Louis Stevenson had been writing the adventure, George might have moved forward, peered down, and gotten a full view of what happened next. But in God's own script, George's fear kept him glued to the helm. By his account, the first thing he knew of what transpired in the cabin was the sound of the axe.

The captain normally slept in his private compartment on the starboard side of the cabin. However, it was a very hot night and, hoping to catch a breeze, he'd rigged himself a hammock, probably across the after part of the compartment where the windows, or *stern lights,* were.

While Humphries held the lantern, Comstock approached Worth's sleeping form and, raising the axe until it bumped the overhead timbers, brought it down with such force that it nearly severed the top of the captain's head from his body.

Comstock then brought his dripping axe to the aid of Payne, who had stationed himself next to Beetle's bunk. The mate slept in a small compartment on the larboard side of the cabin, and the door to the tiny room would have been open because of the heat. As soon as he heard Comstock's killing blow, Payne stabbed through the doorway with his boarding knife. The plan, obviously, was to have

Worth and Beetle killed in their sleep. But the plan went awry.

In George's twisted but appropriate use of butchery terminology, Payne "boned the knife" on Beetle's ribs. The first mate awoke in terror and begged for his life. Comstock would have none of it. "It is a d—d good time to beg now. But you're too late."[5]

As Comstock delivered his death sentence, Beetle was gathering himself. He leapt from the bunk and grabbed the boatsteerer by the throat, knocking him against Humphries, who dropped the lantern. Comstock lost his hatchet in the collision, and was in danger of being choked into unconsciousness by the mate. However, Payne, groping in the dim light cast by the binnacle lamp above, found the weapon and slapped it into his leader's hand. Comstock then hit Beetle over the head and "broke his skull." As the murderer disengaged himself, the mate fell into the pantry where, William Lay reports, "he lay groaning until dispatched by Comstock! The steward held a light at this time, while Oliver put in a blow as often as possible!"

The story of the mutiny on the whaleship *Globe* has been told and retold

229

so many times in the past 175 years that the murders have assumed the stilted, grand, dramatic aspect of men in their extremity, grotesque and removed. Each of the oft-repeated gestures and utterances are the rituals of slaughter. However, on or about 12:15 A.M., January 26, 1824, four killers and their two victims found themselves stuffed into a compartment slick with blood. The space was no more than six feet high and eight feet wide, with the dead captain's hammock, the large dining table, and the companionway from the upper deck taking up most of the available room. Payne's boarding knife, although it was a wicked, deadly tool, was absurdly large for a job that might better have been performed by a straight razor or a sheath knife. Its unwieldy length probably accounted for the botched job, and, in fact, Payne could just as well have "boned" it against the side of the bunk. When Beetle, at five feet, eight inches, leapt from his compartment, the length of his body would nearly have spanned the width of the cabin, knocking the lamp from the steward's hand and sending his surprised assailants into a cursing, sweating, gore-spattered heap. But for Payne's chance proximity to the loose hatchet, the struggle might have gone Beetle's way.

But it did not. He lay in the pantry, gurgling in his own blood and brains.

chapter twenty

The Bloody Hand

After Midnight, January 26

 Lumbert and Fisher, in the compartment just forward of Beetle's, must have waked to the sound of Worth's murder, or to Beetle's plea, before being jolted into full consciousness by the noises of the death struggle. One of them locked the door to the small room from inside. Comstock stationed a man there to guard against the mates' escape and went up on deck to reconnoiter.

George, as he relates in his narrative, was still clinging to his wheel. His brother, all bloody and wild, appeared out of the gloom to light another lantern at the binnacle. Somehow, George found the courage to ask after the fate of Gilbert

Smith, the other boatsteerer. Did Samuel plan on killing him as well?

> He said yes he should kill him and asked where he [Smith] was. I told him I had not seen him (although he had been aft talking with me), for fear if I told the truth he would kill him or go in pursuit of him. He, perceiving me shed tears, asked what I was crying about. I informed him I was afraid they would hurt me. He told me he would if I talked that way.

Samuel went back to his murderous work.

In steerage, immediately on the other side of the bulkhead from the cabin, boatsteerer Gilbert Smith and the coopers Cyrus Hussey and Rowland Coffin had already been wakened by the murders and had made their escape. Hussey and Coffin went to the forecastle. Smith, being in a position of responsibility, ran on deck and ascertained that the disturbance was coming from the captain's cabin. Then, realizing he had no clothes on, he ducked back into steerage to dress. When he returned to the weather deck, he peered down the cabin gangway and saw Comstock with a boarding knife in his hand at the second and third mate's stateroom door.

This was when he approached George at the wheel, who would have confirmed his fears that the officers were being murdered. The knowledge sent him to the forecastle in a hurry. At the other end of the ship, he found the captain's nephew, sixteen-year-old Columbus Worth, who, presumably while on watch, had seen Lilliston going aft with a hatchet earlier in the evening.

At young Worth's urging, Smith hid for a time in a locker "under the heel of the bowsprit." When it was apparent that no one was coming after him, he emerged and warned the men. "I said I reckon they were killing the officers abaft — soon after I heard the report of musquets and the boy Syrus Hussey said he would never see his parents again."[1]

Comstock had gained access to the firearms, which would have been secured in the cabin. He loaded two muskets and went with them to Lumbert and Fisher's stateroom. Lumbert pleaded with Comstock through the locked door and asked if he was going to kill them. Comstock replied in a joking manner, "Oh no, I guess not."

The mutineer then fired through the stateroom door and inquired in the same jocular tone if anyone had been hurt. Fisher cried out that he'd been shot in the mouth. Lumbert, now realizing they'd be shot to death in their compartment, opened the door, still trying to reason with Comstock. As soon as the door opened, Comstock lunged at him with the bayonet of the second musket. Lumbert dodged the thrust, and Comstock's momentum carried him forward into the compartment. The three men wrestled briefly, and as Comstock came up, Fisher had the musket, with its bayonet pressed against Comstock's chest.

As the men confronted one another, Comstock commanded Fisher, in a level tone, to put the gun down, saying he would spare Fisher's life. Fisher obeyed him.

Looking from his compartment into the cabin, Nathaniel Fisher could see Payne, Oliver, and Humphries, all armed and menacing, and he had no idea how many more men were behind them. Comstock's musket shot must have severely demoralized him, and with a bloody mouth full of smashed teeth, he would have had no heart for fighting. His only hope, fed by Comstock's tone, was that the killing was at an end.

He was wrong.

Comstock took the gun and bayoneted Lumbert several times. Then he turned to Fisher and said there was no hope for him, that he too would have to die. He reminded Fisher of his humiliating defeat in the wrestling match aboard the *Enterprise*. By William Lay's account, the stalwart Fisher replied, "If there is no hope, I will at least die like a man!" He turned away and said in a firm voice, "I am ready."

Comstock blew Fisher's brains against the bulkhead. Lumbert, though mortally wounded, begged for water and continued to plead for his life. Comstock responded by stabbing him repeatedly, shouting, "I am a bloody man! I have a bloody hand and I *will* be avenged!"

When he finally came back on deck, the terror had broken. He gave orders to Oliver to call the people up and have them make sail. George was crying in earnest now. Smith, in the forecastle, had realized there was no place to hide. He planned to meet his fate, asking only to be allowed half an hour to perform his religious duties before he died. But Samuel seemed to have sated himself. Calling Smith aft, he asked the terrified boatsteerer if he would support him. Gilbert Smith replied that he would do whatever Comstock said, whereupon the killer clapped him in a bloody embrace — a bizarre action that further traumatized Smith. In attempting to explain it, brother William says, "Smith was a religious young man; and with all his faults, our hero entertained a high respect for sacred things, and a superstitious awe of pious persons." He told George that if he had killed Smith, God would have avenged his death, and that while Smith was with them, "the Almighty would smile on their enterprise for his sake."

Gilbert Smith may have seemed a good luck charm, but it also happened that Smith was the only other man on board who possessed the rudiments of knowledge about navigation. Comstock, who shared quarters in steerage with him, would have known this and perhaps decided it would be useful to have another navigator aboard. If so, it would not be the last instance of intelligence abetting his insanity.

The body of Beetle, in whom, wrote William Lay, "the lamp of life had not entirely gone out," was thrown into the sea through the stern lights. The mutineers disposed of Worth's corpse in the same way, but not before "wantonly piercing his bowels with a boarding knife . . . until the point protruded from his throat."

Payne then ordered William Lay to haul the bodies of Fisher and Lumbert out of the cabin.[2] Third Mate Fisher was disposed of without incident, but Lumbert was still alive. Just as he was sliding over the side, he caught hold of the outermost plank on the vessel's deck and "appealed to Comstock, reminding him of his promise to save him, but in vain; for the monster forced him from his hold, and he fell into the sea. . . ."

Thus, William Lay relates the last moments of John Lumbert's life, but says nothing of his own role in hauling the mate, still alive, up from the cabin, by his ankles, with a rope.

chapter twenty-one

Lumbert's Letter

Pacific Ocean, 1823

Woahoo Ship Globe
Nember the 8 1823

*Most kind and
affectionate parents
it with the grate ist
pleasure that I have an opportunity to
inform you of my health which is good at
present and i ho-ap these few lines will
find you health and sperits with all thay
comforts of life which is all that we can
wish for . . . all tho the pay is small to
beat round cape horn where whales are
wild and scaterind and oil 40 cent
per-galon all tho our — luck small we are
about aleven months out 600 barrels i like
that I have dun my part it is along time
since I heard or seen enny of you which
makes me ankcious to hear from you tell*

sephronia July that thay must write to
me and all the rest of youas i am in north
pasifick ocean a long distance from home
but our ship goes well and we have a
man that will make her walk Capt worth
is a fine man and mr. Beetle is a fine
man and i aminall contented as ever i
was in my life pleas to give my best
respects grandmother and nancy and all
thay rest ned too . . . and so i must
remain your dutifull son

John Lumbert

Lumbert's letter was written in the
Pacific Ocean in November 1823, when
the *Globe* was returning from the Japan
Grounds to Honolulu. At that port, it was
passed on to a whaleship that was homeward bound. By the time his parents had
received it, Lumbert had been dead for
months, though they would not know this
for several months more.

It was the last word from that doomed
ship.

---

* Lumbert's sisters, Sophronia, nineteen, and
Julia, seventeen years old.

chapter twenty-two

The
Fourteen-Second Glass

January 29, 1824

Much to everyone's astonishment, John Lumbert began to swim after he'd been dumped overboard. Comstock ordered a boat lowered after him, but then had a taste of the complexities of command when he realized that the men in the boat might desert to the *Lyra*, which was about two miles off the lee quarter. He immediately reversed his order.

The *Globe* had already raised a lantern, indicating that she was going to tack. At about this time, the *Lyra* answered with her own signal and veered off on her parallel tack, thinking she was following the

241

Globe. However, Captain Comstock ordered the *Globe* to proceed on her original course, thus ensuring that the other ship would sail away from them, and what would soon be Lumbert's corpse.

When day broke he set the crew to work cleaning the carnage out of the cabin. This mopping up of gore was a task for which their whaling experience equipped them, though nothing could have prepared them for what they saw below, "a scene of blood and destruction," says George. "The mates brains were lying in every direction over the cabin floor, the blood of the Capt was likewise strewed over the table and Cabin floor which was cleaned out and every thing brought on deck to be cleaned."

This mirrored the ritual of cleaning up after slaughtering a whale, and it is only the first of several incidents that have this queasy sense of parody. In the same way that the concept of mutiny inverts the normal order of command, so the *Globe* mutiny stood shipboard routine on its head.

Comstock sailed along for a day on a southwesterly course. The crew must have been in shock about what had happened. However, before they could even begin to fathom the hideous events they'd just witnessed, their new master had them back in the cabin preparing cartridges for the ship's fifteen muskets.

The *Globe* was a mutiny ship now, and her crew were outlaws unless they could prove otherwise. Comstock was taking steps to ready himself for whatever violence this new status might entail. But he was also carrying forward the next part of his "plan" to take over a ship, sail to a tropical island, murder the crew, and become a warrior-king in his own paradise.

Sometime that day, in solemn mimicry of Captain Worth's first address to the crew, the mutineer called the men aft and gave them a speech. George recollected, "The principal object was that we should all hang together and obey those who appointed themselves as officers." William Lay had a more specific memory of the occasion. He said that Comstock wrote his rules out and read them like shipping articles before the assemblage.

That if any one saw a sail and did not report it immediately, he should be put to death! If any one refused to fight a ship he should be put to death; and the manner of their death is this — They shall be bound hand and foot and boiled in the *try pots,* of boiling oil.

Everyone was made to sign the articles —
the mutineers in black, the innocent crew in
blue and white.

So, with threats of boiling oil and pots of
colored ink, the new captain assumed
command, some part of him still trapped
in a young boy's fantasy.

On the evening of the next day, January
28, George Comstock went aft to the cabin
on ship's business. Samuel had made him
steward to replace Humphries, who had
been promoted to purser. A purser, in
those days, would most likely be found on
a naval vessel. "He was normally appointed by warrant, and . . . received his
emolument partly by a direct salary and
partly by commission on the issue of daily
victualling allowance."[1] On a whaler, this
victualling would have been performed less
formally by the officers, and perhaps by
the steward and the cook. Humphries had
been granted an empty honorarium, but
the young innocents, George Comstock
and William Lay, reported it matter-of-
factly.

Just as matter-of-factly, Steward Comstock went into the cabin that evening and
discovered Purser Humphries loading a
pistol. "I immediately went to Comstock

and informed him what was going on in the cabin. . . . Coming into the cabin he saw Humphries still standing with the pistol. He inquired of him what he was going to do with it. He told him that he had heard something that he was afraid of his life."

It seems strange that George, who played no part in the mutiny, would alert the head mutineer to a possible insurrection. Comstock and Payne came aft and questioned the purser, who answered in a confused and uncertain manner that Gilbert Smith and Peter Kidder were going to retake the ship, and that he was simply preparing himself for trouble. This answer convinced no one. Comstock's embrace had rendered Smith powerless, and Peter Kidder was, according to George, "a man very easily scared." These two accused men were questioned, and when they denied any plot to take over the ship, Humphries was put in irons and placed under guard.

The next morning, six crewmen armed with muskets brought Humphries on deck and sat him in a chair, guns trained on him all the while, as if he were the lunatic killer. Smith and Kidder were seated on a chest near him. After Comstock assembled a jury of four men,[2] he questioned Humphries again, and the black man, obviously terrified, answered in a confused manner. That, according to William Lay, was the extent of the courtroom drama.

Judge Comstock made a speech, citing the charge against Humphries as the "treacherous and base act in loading a pistol for the purpose of shooting Mr. Payne and myself." He then called upon his jury to render a judgment. If guilty, Humphries would be hanged. If not guilty, Smith and Kidder would hang. The jury found Humphries guilty. Despite the fact that Humphries had been complicit in the mutiny, the fix was in. Comstock and Payne had decided the night before that he should be hanged.[3]

The status of African Americans on whaleships, while better in some ways than it was ashore, was still confused. Some black men rose to be officers. Some ships were manned entirely by blacks. But it was also true that few white whalemen considered black men their equals. African Americans were the stock comic characters of 19th-century whaling narratives.[4] The black, one-eyed sailor in William Comstock's whaling novel is portrayed as superstitious and cowardly. In his biography of

Samuel, William states, "An African is treated like a brute by the officers of their ships."

William's biography is full of hints of Samuel's racial bias. It begins with an odd account of how one of Samuel's ancestors battled and subdued a giant black man in colonial Rhode Island. In the course of his escapades in Valparaiso with John Oliver and the dwarf, Samuel killed a black man who was trying to rob him. These may well be tall tales, but if such stories were passed down as a part of Comstock family tradition, they are evidence of the family's attitude about blacks. William says, bluntly, "Although Samuel Comstock, at the trial, charged the prisoner with a design to shoot Payne and himself, it is not probable that he entertained a very strong belief of his guilt. He was much averse to having a black man on board — he always felt a strong dislike to colored persons."

William Humphries had every reason to be loading pistols. He knew he was a marked man.

Comstock ordered a studdingsail boom rigged eight feet out on the foreyard and a rope rove through a block at its end.[5] Humphries was seated on the rail with a greased rope around his neck and a cap drawn down over his eyes. Each member of the crew was ordered to take hold of the execution rope. When the ship's bell sounded, they were to run aft and haul Humphries aloft.

First, however, Comstock gave the black man fourteen seconds to make a final statement. According to George, "He began — when I was born I did not think I should ever come to this — but the bell struck and he was swung to the foreyard without a kick or a groan."

All the witnesses to Humphries's execution are specific as to the fourteen seconds he was given to make his peace with life. This was the unit of time used for determining the ship's rate of speed with the log and line. Samuel would have been standing near the ship's bell with the *Globe*'s fourteen-second glass in his hand.[6] When the last grain of sand fell through, he struck the bell.

Now each of the men was implicated in the murder of William Humphries. There was no one on the *Globe* without blood on his hands.

chapter twenty-three

Lord Mulgraves Range

February 1824

They lowered Humphries into the sea, but the line fouled in the rigging, and the corpse dragged along in the wake of the ship like a guilty conscience. Comstock ordered a heavy iron blubber hook attached to the body, and when the rope was cut, the body sank from view. Then the mutineers looted Humphries's sea chest and found sixteen dollars in coins that he'd stolen from Captain Worth's cabin after the first round of murders. It was dog-eat-dog aboard the *Globe*. George observed, "Thus it was and always will be with such desperate villains."

The new captain had a thing or two to

learn about sailing a ship. The day after the hanging, he carried away the main topmast backstay by having too much sail up in squally conditions. He steered west from Fanny's Island, then west by north for a day, and then he headed due south for another day, while he set his teenaged crew to work preparing boarding pikes in case another vessel should challenge them. The young men were still very much under his sway. In George's words, "We were obliged to do every thing ordered by our new Capt."

Then, as if he'd finally gotten his bearings, Samuel Comstock kept the ship headed west by south for a week. During this interval, a kind of postrevolutionary calm seems to have prevailed. A few casks of oil and some deck gear were jettisoned, but not in any organized manner. Stephen Kidder told Consul Hogan, "The men were painting the ship black, defacing the name of the ship from the stern & wherever it was branded on the spars, oars, buckets &ct." Comstock was moving forward with his ultimate plan to remove every trace of the *Globe*.

Captain Comstock also invited all of the crew aft to share their meals in the cabin. This invasion of officer's country by the foremast hands symbolized the mutiny's final upset of shipboard order. Dinners there must have been strange affairs indeed: rum flowing and the officers' mess pillaged for delicacies, while behind the debauchery lingered images of the slaughter that had just taken place.

William Comstock reports that on one of these occasions in the cabin, Oliver and Payne complained of being haunted by dreams of the violent murders. "Our hero laughed at their terrors, and told them that the captain had also appeared to him, and shook his gory locks and pointed at his bloody head — 'but,' said Comstock, 'I told him to go away, and if he ever appeared again, I would kill him a second time!' "

Comstock remained alert for chances to advance his plan. Anthony Hansen, the Indian from Cape Cod, later testified he'd overheard the mutineer planning to set some people adrift.

That opportunity never materialized, however, and on February 7, 1824, the *Globe* approached a part of the Gilbert Islands known in those days as the Kingsmills Group.[1] Most of the islands in this chain were coral atolls, the sunken rims of ancient volcanoes. Typically, these atolls took the form of a ring of islets surrounding a central lagoon.

The narratives of William Lay and George Comstock both give an abbreviated journal of the *Globe*'s sail to the Kingsmills, including a daily record of their longitude. Since it is unlikely that they had a chronometer aboard or that Samuel was sufficiently versed in navigation to perform the complex calculations necessary to take lunars, it seems they were keeping very careful track of their course by dead reckoning. This makes it clear that Samuel hadn't just stumbled across these islands. They were a destination. Samuel would have cast about for a refuge after the mutiny and, knowing that he couldn't risk the relatively crowded shipping lanes, he would have recalled the Kingsmills. He'd visited them during his whaling voyage aboard the *Foster*, and they were less than two weeks' sail from Fanning Island.

George's account refers to the islands as "The Kingsmills Groupe," which corresponds with the way the name was spelled on a popular English chart of that era.[2] It is likely that one of these charts was aboard the *Globe* and that Samuel used it to navigate. His plan to be king of a tropical island was proceeding under its own cracked logic.

The Kingsmills are part of a 1,500-mile string of islands composed of the Gilberts and the Marshalls. This long chain runs in a northwest to southeast direction across the equator and is located midway between Hawaii and New Guinea. Though first touched by the Spanish and the Portuguese in the 16th century, the islands in this chain were little frequented by westerners, since most of Europe's trade with Asia moved via the Cape of Good Hope. Bougainville, Lapérouse, and Cook, the great navigators of the 18th century, missed them, and it wasn't until the late 1700s that the English began taking an interest in this distant outpost of the Pacific.

In 1788, two English merchant captains explored the islands systematically and made them known to the world. Their names were Gilbert and Marshall, and so the two island groups in that chain became known. Gilbert and Marshall had been hired by the East India Company to carry the first batch of convicts to Botany Bay. On their return from this historic jail-run, they sailed for Canton, and it was on this leg of the journey that their discoveries were made.[3]

Gilbert published a book about his discovery, and subsequently the islands were visited by British and American merchantmen seeking new trade routes to China and, less frequently, by whalemen (whaling routes were just beginning to extend this far). However, only one other formal expedition preceded the arrival of the *Globe*. This was led in 1817 by the Russian explorer Kotzebue, who'd been sent out by the chancellor of the empire to explore the Pacific. Among his company were an author-naturalist named Chamisso, and a marvelous artist named Choris. Each of these men wrote accounts of the expedition just a few years before Comstock's arrival.

There were still many islands whose inhabitants had never seen a white man. The people Comstock and his crew met were probably similar in their manners and customs to the natives Kotzebue had met seven years before. Both parties would have encountered people whose ways had yet to be changed by westerners. Otto von Kotzebue described them as being "tall and well shaped; the naturally dark colour of their bodies appeared black at a small distance, because they are tattooed; they do not tattoo their faces. . . . They rub their long black hair with cocoa oil, tie it together above the forehead and adorn it with flowers and wreaths made of shells . . . some had a couple of fine matts tied round the body; others wore a braided belt from which the fibres of grass hung down to the feet. We were most struck by their ear-holes [in the lobes], which measured more than three inches in diameter." He found them an attractive people. "A high forehead, an aquiline nose, and sparkling brown eyes, advantageously differentiate the natives . . . from the rest of the South Sea islanders."[4]

On February 7, two weeks after the mutiny, the *Globe* made her first landfall at the northern end of the Gilbert Islands. Probably she landed on or near the large atoll now known as Tarawa, which was the site of a famous battle against the Japanese in World War II.[5]

Comstock's men met with a less friendly reception than the Russians had seven years before. George said the natives appeared "quite hostile." Perhaps this was just a local difference of temperament (they were still several hundred miles south of where the Russians had landed), or the result of some bad experiences in the intervening years. Comstock sailed the *Globe* a little farther north to another island.

One of the whaleboats went ashore to reconnoiter and trade with the natives, but this proved to be impossible. Before the whaleboat could land, the men from the island swam out to it, heaved themselves aboard, and attempted to steal whatever they could lay their hands on. The crew of the whaleboat sent them over the side and discharged a volley of gunfire, which frightened the islanders and may have wounded some. Then the boat raised its sail and set off downwind after a small canoe paddled by two natives.[6] When they drew within range, they fired into the canoe and hit one of the men. As the mortally wounded native slumped down in the canoe, his terrified companion offered their clothes and beads in exchange for their lives. William Lay, who presumably watched from aboard the *Globe*, was appalled by the treatment of these innocents. "Here was another sacrifice; an innocent child of nature shot down, merely to gratify the most wanton and unprovoked cruelty . . . and when the years have rolled away, this act of cruelty will be remembered by these Islanders, and made the pretext to slaughter every white man who may fall into their hands."

This earnest reflection prefigured one aspect of relations between white men and Pacific islanders for the next several decades. The whaleboat returned to the ship, and Comstock, having had enough of the Kingsmills, headed north for Lord Mulgraves Range.

This island, now known as Mili Atoll, is at the southern end of the Marshall Islands. When Captain Marshall first sighted it in June of 1788, he named it after Lord Mulgrave, who was a celebrated captain in the Royal Navy. It was typical in its configuration, being a ring of islets, some only a few hundred yards wide, surrounding a central lagoon about twenty-five miles across. From water level, the circular nature of the atoll is not apparent, and the smaller islands composing it appear to be grouped or ranged in clusters on the horizon. The group of islets was also referred to as Mulgrave Islands and so it is named on that English chart, which was probably Samuel's navigational aid. When the Kingsmills failed to provide the hospitality Comstock sought, the Mulgrave

Islands — Mili Atoll — became his destination.

The journey from the Gilberts to Mili Atoll was about 300 miles. It took the better part of three days, during which the *Globe* experienced headwinds and squalls.[7] On February 11, they saw the Mili off to the northeast and stood in close to shore. Their reception here was much better. The expected fleet of canoes came out to the ship, and this time they were able to trade for food without being robbed. Better still, the canoes carried women and girls. With no Captain Worth to disapprove, the men brought them aboard for the night.[8]

The next morning, Comstock and his crew returned the women ashore and explored some of the small islets in the atoll. The mutineers had hopes that the soil would be suitable for raising crops, but in this they were disappointed. Still, Samuel Comstock seems to have decided that Mili met his requirements in other respects, and he continued exploring for two more days in hopes of finding a decent harbor. Finally, on the thirteenth of February he settled on an anchorage a few hundred feet off "a low, narrow island."

Ironically, the waters between the

Gilberts and Mili Atoll swarmed with sperm whales. Many were sighted, but the boats were never lowered. The *Globe* was on a different journey now.[9]

chapter twenty-four

The Man of Blood

Mili Atoll, February 17, 1824

There the *Globe* lay, fifty yards offshore, on the precipitous edge of a coral atoll, itself the tip of an ancient volcanic mountain. Her stern was kedged seaward, her bow anchored in seven fathoms of water. And there, at an equally precipitous place in the story, George Comstock's handwritten narrative ends. It seems an odd place for him to have stopped; clearly, George did not share his brother's flair for drama.

In another context, however, it is perfectly understandable why his narrative ends where it does. George Comstock drafted this account at the behest of Gorham Coffin and the Mitchell partners. The owners were highly motivated to discover all the facts about how the mutiny had occurred and who was responsible. Beyond that, their official interest waned. By the time the *Globe* reached Mili Atoll, everything that was of importance to them, in terms of assigning responsibility, had already happened.

They anchored in the southwest corner of the atoll, along the southern arm of its largest island, Mili-Mili. To the north, the island widened into a dense forest of coconut palms that sheltered a native village. To the east, it extended as a narrow strip — lagoon on one side, ocean on the other, jungle in the middle — for another thirty miles in a series of elongated islets including Arbar, Enajet, and Lukunor. On Arbar, the islet just east of them, was another village. All these islets were connected by sand spits (the men called them *causeways*), making it possible to cross from one islet to the next, all the way from Mili-Mili to Lukunor.

Comstock had his crew build a floating dock of spare yards and lumber from the ship, with one end resting upon the rocky beach and the other end, like the *Globe*, held seaward by an anchor. Onto this platform they unloaded boatloads of supplies that had been ferried to shore. Then the mutineers constructed a larger landing craft by lashing more spars across two boats. This contraption could be run on shore at high water, substantially speeding up the process of looting the ship. William Lay gives a detailed list of the items they took out of her.

One mainsail, one foresail, one mizzentopsail, one spanker, one driver, one maintop gallantsail, two lower studdingsails, two royals, two topmast studdingsails, two top-gallant-studdingsails, one fly-gib (thrown overboard, because a little torn) three boat's sails (new) three or four casks of bread, eight or ten barrels of flour, forty barrels of beef and pork, three or more 60 gal. Casks of molasses, one and a half barrels of sugar, one barrel dried apples, one cask vinegar, two casks of rum, one or two barrels domestic coffee, one keg W.I. coffee, one and a half chests of tea, one barrel of pickles, one do. Cranberries, one box chocolate, one cask of towlines, three or more coils of cordage, one coil rattling, one do. lance warp, ten or fifteen balls spunyarn, one do. worming, one stream cable, one larboard bower anchor, all the spare spars, every chest of clothing, most of the ship's tools, etc. The ship by this time was considerably unrigged.

Just exactly how "unrigged" the ship would have been can be roughly calculated by assuming that the *Globe*, like most whalers, carried two sets of sails. After they stripped her, therefore, she would at least have been left with one course each on the fore- and mainmasts, topsails for all three masts, a topgallant sail for the mainmast, headsails, a spanker, and a driver. This was more than enough to keep her seaworthy, and it suggests that the ever-calculating Samuel wasn't committing himself to the destruction of the ship.

The sails and spars would have gone ashore to make a tent for shelter until some permanent structures could be built. The bread, flour, beef, pork, and molasses would have been their staple provisions, and the officer's dainties — the cranberries, sugar, tea, chocolate — might have done a bit to sweeten the harsh exile of mutineer and innocent crewman alike. They spent one peaceful day pillaging the *Globe*. Then, on the sixteenth of February, three weeks after the night of the mutiny, trouble started.

Silas Payne, Comstock's first lieutenant, was on the ship selecting articles to be sent ashore and supervising the loading of the landing craft they'd improvised. Comstock, Smith, and a number of other men were on the island, unloading the goods as they came in. Undoubtedly, Payne assumed he was selecting articles that would be used by the crew to establish their colony on the island. He must have been surprised to see Comstock giving these valuable supplies to the natives. Surprise would have grown into shock as he watched this practice continue, then into rage as he realized what Comstock had in mind.

The head mutineer had elected to be the man on the beach for a very specific reason. By liberally bestowing gifts on the natives, he was winning their allegiance. They'd know he was the chief, and when the time came, they'd be ready to do his bidding. Whether or not Payne grasped the whole of Comstock's plan at first, the fact is that he saw his leader acting contrary to the group's interest, and he understood that no good could come of it.

William Lay suggests that Comstock was befriending the natives in order to defend himself against an uprising by the innocent

crew members. There may have been mur-
murings to that effect among the innocent
crew, but this was still a band of demoralized young men. Comstock would not
have seen them as a threat.

Rather than use the natives to defend
himself against the likes of William Lay, it
seems more probable that Samuel Comstock sought their allegiance in order to
enhance his power over the entire crew —
Payne included. As the hours went by, and
the lanky assistant mutineer laboriously
reconstructed his leader's thought pro-
cesses, he would have come to understand
that the natives were the strategic key to
the situation on Mili Atoll. The man who
controlled them controlled the destiny of
everyone on the *Globe*. Payne saw that
Comstock had moved to do this without
consulting him, and he began to see that
Comstock was not finished plotting.

Payne had been an eager accessory to
the murders aboard the *Globe*. He might
have been as debased as his leader; he certainly had as little to lose. Once he decided
it was necessary to challenge Comstock,
there was no way the situation could end
in reasoned negotiation aimed at address-
ing the best interests of both parties. At
this point the tattooed islanders were no

more than pawns.

Silas Payne sent word to Comstock.

William Lay, who might have acted as courier, recalled the substance of the message, "That if he did not act differently with regard to the plunder, such as making presents to the natives of the officers' fine clothing etc., he would do no more, but quit the ship and come on shore."

After receiving Payne's message, Comstock called him ashore, and the two men had a long and heated argument in the tent. William Lay says the battle of words came to a head when Comstock raged, "I helped to take the ship and I have navigated her to this place. I also have done all I could to get the sails and rigging on shore, and now you may do what you please with her; but if any man wants any thing of *me*, I'll take a musket with him!"

Comstock's attention had wandered. He'd ceded too much. Payne instinctively understood that if he controlled the ship, he would be back in command of his own fate. Sensing a weakness, he challenged his chief for leadership.

Silas Payne replied, "That is what I want, and am ready."

This confident stance was the last thing Comstock expected. The charismatic and inventive killer, who'd never been at a loss, was suddenly nonplussed. He sidestepped Payne's invitation to duel, saying that he only wished to return to the ship, after which Payne could do whatever he wished.

Back aboard the *Globe*, Comstock called for the "laws" he'd written out and, before the astonished crew, cut them up with his sword, announcing that he would no longer be their leader. Brother William paints an almost touching portrait of the killer fuming and blustering before a passive audience. "Thinking the hands on board appeared pleased by his desertion of them, he became enraged and challenged them to fight him." They did not respond, and he stormed off the ship. "I am going to leave you — Look out for yourselves." In all probability they were only too happy to do so.

Comstock had Payne row him back to shore, and before the boat even touched the beach, the furious mutineer jumped into the water and waded to the tent, where he went to the captain's chest and provided himself with fishing gear and a jackknife. He then headed eastward down the narrow island, in the direction of Arbar.

As William Lay tells it, "He took his departure, and as was afterwards ascertained, immediately joined a gang of natives, and endeavored to excite them to slay Payne and his companions! At dusk that day he passed the tent, accompanied by about 50 of the natives, in a direction of their village." The mutineer's plan was completely exposed now.

Payne went back aboard ship and explained that Comstock wanted to persuade the natives to kill them all. He ordered some of the crew to accompany him ashore, where they spent the night guarding the tent and their supplies. Although Payne had succeeded in deposing Comstock, he found himself in an awkward position strategically. It would have been best simply to sail to another island and leave Comstock alone with his fishhooks and natives. But a good portion of the ship's supplies and half her sails had already been unloaded. Payne and the others were forced to remain ashore to protect their resources.

During what must have been a long night, Payne asked Gilbert Smith to help him murder Comstock. But, says Lay, "Poor Smith, like ourselves, dare do no other than remain upon the side of neutrality." The deeply religious boatsteerer may not have wanted to murder anyone — even Comstock. But the other young crew members, still dominated physically and psychologically by the older mutineers, thought it better to let them battle one another and not become involved.

Next morning, the men in the tent spotted Comstock some distance away, moving briskly toward them. Payne's side-kick, John Oliver, had been keeping watch from the ship. As soon as he saw Comstock, he rushed ashore to join Payne, and probably Thomas Lilliston, the coconspirator who'd refused to take part in the murders of Captain Worth and the mates. These men were joined by a fourth, most likely Joseph Thomas, who was on the fringe of the conspiracy. They loaded their guns and waited for the arrival of their former chief.

Samuel drew his sword and approached "in a menacing manner." But as soon as he saw the muskets aimed at him, he waved his hands and told them not to shoot, that he wouldn't hurt them. No doubt recalling a similar promise made to Nathaniel Fisher, they fired, hitting him twice. According to William, "One struck him on the upper lip and passed through his head;

the other entered his right breast and passed out near the back bone. As soon as he had fallen, Payne ran to him with an axe, and buried it in his brains."

Muskets of the early 19th century were not rifled like modern guns. They fired round balls with no stabilizing spin and were notoriously inaccurate. The fact that Samuel B. Comstock was hit by two of the four untrained shooters suggests he was virtually on top of them when they fired.

Silas Payne, after using a hatchet to certify the death of his former chief, proceeded to inter Comstock with not-quite-full military honors — the final parody in the series of such events that characterized the mutiny. Lay commented upon it. "Every article attached to him, including his cutlass, was buried with him, except his watch; and the ceremonies consisted in reading a chapter from the bible over him, and firing a musket."

They read the Bible, they fired the salute, but they kept the watch.

Samuel B. Comstock was dead, a little more than three weeks into his reign. In this brief interval he had displayed a malevolent savagery unparalleled in the annals of American maritime history. He had also demonstrated leadership, cunning, bravado, and, in his last hours, despair. After 175 years, the desire to understand this strange, twisted character lingers over the story like gun smoke from his final salute.

Unfortunately, 19th-century storytellers and their audiences did not require such analyses. Comstock's trumped-up grievances against the *Globe*'s officers were duly listed, and deeper questions of motivation were dismissed with pat phrases. To Lay and Hussey, Comstock was simply a force of evil, a "monster." Samuel's father saw him as one who had fallen from grace. When told of his son's misdeeds, Nathan Comstock lamented, "Oh . . . heaven-forsaken Samuel."[1] William Comstock, while arguing that his brother's acts were the premeditated culmination of a lifelong fantasy,[2] repeatedly hinted that his brother was possessed by a kind of insanity. His description of Samuel in the midst of his killing frenzy perfectly coincides with the Romantic era's image of the maniac. "His face, arms and breast were bloody — his eyes flaming with fury — and his shirt nearly torn from his back."

In his introduction to a 1963 reprint of Lay and Hussey's narrative, Edouard

Stackpole tentatively addressed the modern reader's need for deeper explanations. "As for the most controversial figure of all — Samuel Comstock, the harpooner turned murderer — here is a study for the psychiatrists." Stackpole, the man who first assembled and sifted through the documentation regarding the *Globe* mutiny, never followed up on his own suggestion.

Dr. James McGee is a forensic psychologist of wide experience and vast portfolio.[3] In the spirit of intellectual adventure, he agreed to examine Comstock's life and crimes and venture an analysis of the mutineer. He cautioned that the evidence was extremely limited, and the conclusions would only be educated guesses. Still, what he had to say surprised me.

In that Comstock had sufficient mental capacity to plan, to be aware that his actions were wrong, and to have changed them, no court would find him unfit to stand trial or innocent by reason of insanity. He did, however, show marked evidence of personality disorder — what people in Dr. McGee's line of work referred to as "Mixed Personality Disorder." This condition combines features of Narcissistic, Antisocial, and Paranoid disorders. McGee characterized personality disorders as patterns of behavior that drive everyone else crazy. They're ultimately self-defeating, but while they're happening they're much more annoying to others than to the person exhibiting them.

Typically, such disorders come on in adolescence. In young Samuel's case, Narcissism — grandiosity, self-importance, and fantasies of success, coupled with a lack of empathy — was a good fit with his biography. People with Paranoid Disorder persistently bear grudges and misinterpret the actions of others in a negative way. That fit another portion of the biography — the wrestling match with the third mate, for example, or his irrational dislike of the captain. As for the Antisocial component, Comstock had a marked disregard for the rights of others. He lied, used people, and he showed no remorse.

Additionally, there was a hypomanic element to Comstock's behavior. He demonstrated enhanced creativity (albeit perversely so) and persistent, intense, elevated activity lasting several days. According to Dr. McGee, successful businessmen are sometimes hypomanic — extremely focused, able to go without sleep, very goal oriented. It could be a good way to get things accomplished if it didn't get out of control. But when hypomania is coupled with certain personality disorders, it makes for a dangerous combination, often violent.[4] Many criminals exhibit this mix of hypomania and personality disorder. Charles Manson, for example, may have. In fact, Dr. McGee saw Samuel Comstock as a sort of 19th-century Manson. Neither man was legally insane at the time of the crimes, but both exhibited signs of personality disorder and hypomania. Dr. James McGee believed that if he had survived, Samuel Comstock would have progressed into full-blown paranoid schizophrenia.

As for causes, there were only theories. "The thing to remember," he said, "is that real life is complex. Behaviors are often multiply determined. Any time I read a really tight explanation of something like this, I reject it out of hand."

With that, the specter of "the Man of Blood" again shouldered his mantle of enigma.

chapter twenty-five

The Great Escape

Mili Atoll, 1824

On his way to the guillotine in 1793, Louis XVI supposedly asked his executioners, "Is there any news of Lapérouse?" The story was probably intended to illustrate the monarch's frivolity, but it also serves as an indication of the importance the French placed on the fate of their renowned navigator. As American traders began to find their way into the Pacific, the wealthier governments of England and France were sending official expeditions to explore and develop trade in this vast region. Cook, in his three expeditions in the 1770s, made many important discoveries for England. In 1785, the French sent Jean-François de Galaup Lapérouse with two frigates to visit waters overlooked by Cook, in hopes of finding a northwest passage. He was also to bring back information on the whaling industry, the fur trade, and Spanish California, and he was to survey the Pacific with an eye toward expanding French commerce.

Although he never discovered a shortcut to the riches of the Orient, Lapérouse accomplished a good part of this ambitious agenda. He sent dispatches to France from Kamchatka and Botany Bay, and these writings became the basis of a four-volume illustrated report on his expedition. His efforts enabled French geographers to correct the charts made by other nations, and his far-ranging travels left a lasting French presence in many parts of the Pacific. On the final leg of his journey, misfortune overshadowed his substantial accomplishments; in 1788, he left Botany Bay in Australia and was never seen again.

Like the tragic death of Cook a decade before, the utter disappearance of Lapérouse captured the world's imagination and focused it once again on the mystery and possibility of this mighty ocean.[1] The expeditions of D'Entrecasteaux and Dumont d'Urville were at least partly search missions, but they were unable to find any definite evidence of Lapérouse's whereabouts. Finally, in 1827, a contentious but savvy English South Seas trader named Peter Dillon discovered the remains of the expedition.

Dillon had stumbled onto some of the expedition's artifacts in 1826 while trading sandalwood for the East India Company. He returned to his home port in Bengal and convinced the government of British India to finance an expedition under his command to go back and determine once and for all what had happened to Lapérouse. After a series of adventures and mishaps in Australia, New Zealand, and Tonga, he discovered the wreckage of Lapérouse's two ships off the Santa Cruz Islands, north of the New Hebrides (now known as Vanuatu). Peter Dillon took his discovery to France, where Charles X conferred upon him the order of the Légion d'Honneur and an annuity of 4,000 francs. He wrote a book about his adventures and established himself as a minor but fascinating character in the ranks of that era's explorers.[2]

Dillon was an irrepressible storyteller who couldn't resist larding his book with yarns and anecdotes. One of these side-stories took place in June 1824, five months after the *Globe* mutiny, when the author and his ship, the *Calder*, were anchored in the South American port of Valparaiso on another sandalwood voyage. From this vantage point he saw a strange ship approach with her ensign flying upside down in the universal signal of distress.[3]

The situation on Mili Atoll in February 1824 had gone from bad to worse. Silas Payne may have had the physical courage to challenge Comstock, but he did not possess Comstock's leadership abilities. Both he and the Englishman John Oliver fell into a sort of despair after murdering their leader. As deranged as he may have been, Samuel Comstock had provided the illusion of authority. His crazy plan had given them a semblance of structure. Now the two junior mutineers were faced with the reality that they had nominal control over a ship that they didn't know how to navigate, and that they were dependent on a crew who would desert them at the earliest opportunity. In addition, there was the ever-present fact that they had committed an offense for which they would be hanged if they ever returned to civilization. That day, Lay says, "No duty was done." William Comstock enlarges on their malaise, bringing back the image of his brother, "The magical charm in which he had bound their sense was broken. The false enthusiasm which he had kindled, had sunk with him in his grave, like a candle expiring in the socket."

Finally, toward evening, Payne roused himself and ordered a detail of men back to the *Globe*. Joseph Thomas, in later testimony, said they were sent aboard to get supplies. William Lay said Payne simply wanted the men to mind the ship. Whatever his reasons, this was the first of Silas Payne's bad decisions as de facto commander of the *Globe*'s company.

By this time, there was a plot afoot to retake the ship. Peter Kidder would later claim it had first been discussed when the *Globe* was on her way to the Kingsmills. Gilbert Smith said it was an idea he had shared with George Comstock and the Kidder brothers. Even the suspected mutineer, Joseph Thomas, said that George had approached him about it on the day Comstock was killed. Most likely, every honest member of the crew was aware of it, though the level of fear and paranoia must have been such that each man believed the secret was known only to a few.

How concrete a plan could Smith and the others have put together? Comstock and his lieutenants held the muskets and the psychological dominance. Otherwise, the situation was completely fluid. Gilbert Smith and his mates were probably just waiting for an opportunity. With Comstock's murder the chance presented itself, and in Payne's dim-witted order, sending a crew of honest men aboard the ship, it landed in their laps.

Stephen Kidder, who was sick, had remained aboard the *Globe*. Payne detailed Gilbert Smith, Joseph Thomas, George Comstock, Anthony Hansen, and the Kanaka known as Joe Brown to go out to the ship.[4] If Payne thought that Joseph Thomas would keep these men in line, he was wrong.

Smith signaled ashore for Peter Kidder to join them. Kidder went to Payne and asked for permission to visit his sick brother. Payne said he could go if one of the other men would stay ashore in his place. Peter Kidder asked this favor of Joe Brown, and the Hawaiian agreed to let Peter replace him on the ship detail.

Plans had been made for some of the others on shore to swim out to the *Globe* under cover of darkness, but this was not

280

to be. Smith and his crew stealthily loosed the fore and main topsails and readied the ship for departure, but by this time the moon had begun to rise. It would be too dangerous for them to wait any longer.

At 9:30 on the evening of February 17, Gilbert Smith set to work on the hawser with a greased saw that he'd laid aside, and then gave the order to let fall the sails. In two minutes the cable had parted. The ship payed off, and when her bow was seaward and the offshore breeze had filled her sails, he cut the mooring line and they were away. As painful as it may have been, they'd done the right thing not waiting for their comrades to swim to the ship. According to William Lay, Payne had set two men on watch, presumably to guard against the natives. Shortly after the moon rose, one of them looked seaward and immediately raised the alarm, "The ship has gone! The ship has gone!"

She was gone, all right, but where she'd wind up was anyone's guess. The odds did not favor six young men in a full-rigged ship with limited supplies, few navigational aids, and 7,000 miles of Pacific Ocean to cross.

The ship that Peter Dillon spotted

coming into Valparaiso was, of course, the *Globe*. Somehow Smith and his men had brought her from Mili Atoll to within thirty miles of their intended destination.[5] They'd been at sea for 109 days.

The bedraggled ship came to anchor, the young crew standing on her cluttered deck, equally bedraggled, slightly hysterical with relief at being safe in port, and inarticulate, initially, because of the tremendous pressure of the events behind them. Dillon was probably making his way from one crew member to the next, already homing in on the juiciest parts of the yarn. Behind him, the American consul Michael Hogan would have been doing more listening than talking.

It was an epic journey, and yet it is the only portion of the entire *Globe* saga for which there is no narrative account. Presumably, the events Smith and his men were fleeing were of such immensity that, when it came time to tell their tales, their three-month voyage seemed only an afterthought. William Comstock, who had his information from George, says only that it was a "long and boisterous passage." Gilbert Smith didn't have much more to report. "In a distressed disabled ship we made the best of our way as winds and weather would permit for this port . . . [we] had no quadrant on board but intended to stand by the wind till we got into the variables and to steer to the eastward to make some part of this coast."

According to both Lay and Smith, they had charts and a compass aboard, but without a chronometer to assist in ascertaining their longitude or a quadrant, which was used in determining both latitude and longitude,[6] they could only estimate their position.

This is important because it answers Edouard Stackpole's later question "as to why Gilbert Smith did not lay a course for the Sandwich [Hawaiian] Islands that were immeasurably closer than the coast of South America."[7] The answer is that they were inexperienced navigators, sailing nearly blind. If they did escape with the ship, they weren't at all certain they could find the Sandwich Islands, whereas if they headed east and didn't sink or starve, they'd eventually reach South America.

Much of the western coast of South America was embroiled in a series of revolutions against Spanish colonial authority. In many ports, the reception accorded an American ship depended on who was in power that week or month or year. Smith was not only escaping from the Mulgraves. He was also going to get help for the companions he'd left behind. This meant contacting American officials, and as far as most whalemen were concerned, Valparaiso would have been a logical choice. Other ports, such as Callao, far to the north, might have been equally good destinations, but Smith's experience of South American ports was probably limited to what he'd learned on the *Globe*. Valparaiso was the only port on that coast at which she had stopped.

For all the off-loading of supplies in the Mulgraves, the mutineers had left a sufficiency of beef and bread aboard. As far as potable water was concerned, when the *Globe* left Oahu in January, she would, as a matter of standard practice, have carried enough to last her crew several months. Less than thirty days after departing Honolulu, they were on Mili Atoll, probably with two-thirds of their water in reserve. Since water was available on the atoll, the ship's supply would've been left on board. A two-month supply for twenty-one men would easily have carried Smith and his five companions through their 109-day journey.

One thing the survivors did report to

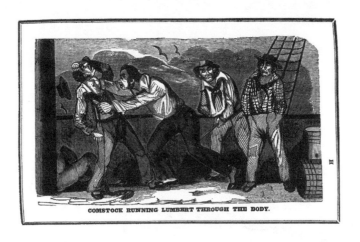

COMSTOCK RUNNING LUMBERT THROUGH THE BODY.

THE EXECUTION OF HUMPHRIES.

THE DEATH OF SAMUEL COMSTOCK.

Engravings courtesy of the
Peabody Essex Museum

285

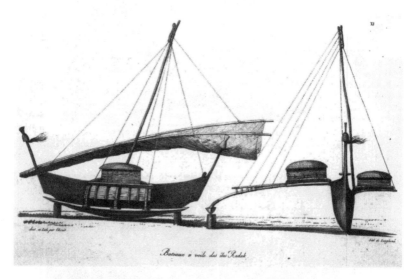

Marshall Islands canoe, circa 1820

Scene on the Marshall Islands, circa 1820
Peabody Essex Museum *for both pictures.*

Portrait of Captain Thomas Worth
Courtesy of the Martha's Vineyard
Historical Society.

The Whaleship Globe
Sail and Rigging Plan

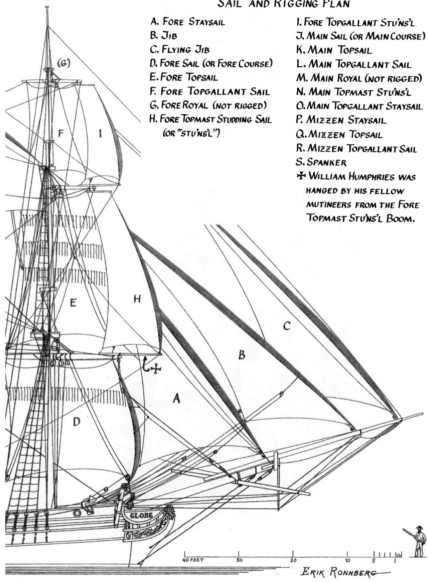

A. Fore Staysail
B. Jib
C. Flying Jib
D. Fore Sail (or Fore Course)
E. Fore Topsail
F. Fore Topgallant Sail
G. Fore Royal (not rigged)
H. Fore Topmast Studding Sail
　(or "Stu'ns'l")

I. Fore Topgallant Stu'ns'l
J. Main Sail (or Main Course)
K. Main Topsail
L. Main Topgallant Sail
M. Main Royal (not rigged)
N. Main Topmast Stu'ns'l
O. Main Topgallant Staysail
P. Mizzen Staysail
Q. Mizzen Topsail
R. Mizzen Topgallant Sail
S. Spanker
✠ William Humphries was hanged by his fellow mutineers from the Fore Topmast Stu'ns'l Boom.

GLOBE

40 FEET　30　20　10　1

Erik Ronnberg

XIII

Homme des îles Radak.

Man of the Marshall Islands, circa 1820
Peabody Essex Museum

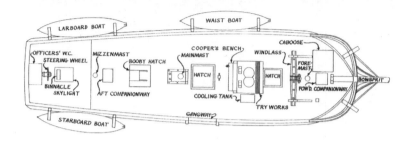

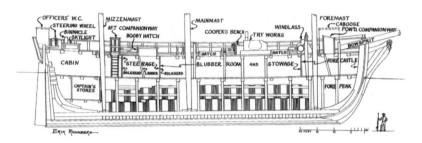

ERIK RONNBERG

SHORTLY AFTER MIDNIGHT, JANUARY 26TH, 1824, SAMUEL COMSTOCK ROUSED THE STEWARD, HUMPHRIES, AND GOT WEAPONS FROM THE STEERAGE COMPARTMENT.

THEY MET SEAMEN PAYNE, OLIVER AND LILLISTON WHO HAD COME FROM THE CREW'S QUARTERS IN THE FORECASTLE, AND THE FIVE MEN WENT AFT TO THE CABIN COMPANIONWAY, WHERE LILLISTON DESERTED THEM.

COMSTOCK, PAYNE, OLIVER AND HUMPHRIES WENT BELOW INTO THE CABIN AND MURDERED CAPTAIN WORTH IN HIS HAMMOCK.

PAYNE WOUNDED FIRST MATE BEETLE IN THE 1ST MATE'S CABIN. BEETLE STRUGGLED WITH THE MUTINEERS UNTIL COMSTOCK MURDERED HIM IN THE PANTRY.

COMSTOCK THEN STABBED SECOND MATE LUMBERT AND SHOT THIRD MATE FISHER IN THE 2ND AND 3RD MATES' CABIN.

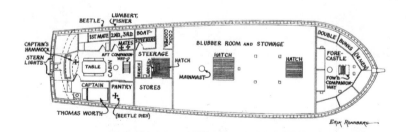

ERIK RONNBERG

291

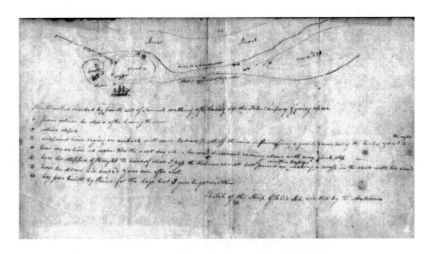

George Comstock's map
Nantucket Historical Association

List of Globe *crew members*
Courtesy of Rennie Stackpole

Consul Hogan was that Joseph Thomas, the suspected mutineer, was not much of a shipmate during their escape to Valparaiso. Gilbert Smith said, "When help was called he was generally the last man. Disobedient & often said he would do as he thought fit." Gilbert Smith's crew thus consisted of Stephen and Peter Kidder, George Comstock, and Anthony Hansen, with Thomas playing only a grudging role. Assuming they were neither hungry nor thirsty, there is still the question of how five or six men could have managed a ship that was normally run by a crew of twenty-one.

Significantly, no topsails — the main working sails of a square-rigged ship — were removed from the ship at Mili. One spanker, or *steering sail,* had been offloaded. This was the gaff-rigged sail lowest on the mizzenmast, and it was used to help maneuver the ship. But the mizzen staysail, which was not taken ashore, is far enough aft that it could have been used to help steer the vessel. Additionally, most whaleships carried two suits of sails. The second set (probably old, worn sails, or those reserved for heavy weather, such as would be encountered rounding Cape Horn) could well have been left aboard. If this were the case, there would have been plenty of canvas.

In a pinch, four people can furl a topsail.[8] No matter what the sail plan, there would be one man to steer and at least four to handle the sails. And, in a life-and-death situation even the recalcitrant Joseph Thomas would have "turned to." Probably the most grueling aspect of the voyage was that the crew would have stood *watch and watch* — four hours on and four hours off, in shifts of three men each — for three months, with all hands called up when some alteration in the sail plan was required.

The literature of the sea provides many accounts of successful shorthanded voyages. The *Globe* completed her voyage at an average of not quite seventy miles per day — slower than three miles per hour. This was a leisurely pace, even for a whaleship. It suggests intelligent and conservative sail handling. In the open ocean, the *Globe* was at greatest risk when caught in deteriorating weather with too much sail up. So it is likely that Gilbert Smith was content to poke along, even in fair weather, under just as much sail as his crew could manage — topsails and perhaps a jib or staysail.

Weather was also a critical factor. February is still the stormy season in the Marshall Islands. When traveling from the Gilbert Islands to Mili, Comstock and his men had sailed into wind and squalls for three days. These same northerly winds must have been in play on the night of Smith's escape, since the breeze blew them off the southern shore of Mili Atoll. If it persisted for another few days, they would have been well on their way through the Gilberts.

Although none of the crew made any comment regarding the weather during their escape, Chevalier Peter Dillon had an observation to add. The anecdote about the *Globe* is included in Dillon's book not just because of the dramatic value of the mutiny and the escape, but by way of proving one of Dillon's pet theories. Over the years, he claims to have observed a seasonal northwesterly wind in certain areas of the South Pacific. He tells of a native canoe blown hundreds of miles from the Tonga Islands to the Navigators Islands (now called Samoa) on this steady wind. He adds the story of the *Globe* and her voyage as further proof of his hypothesis about this seasonal wind. When Smith and his companions realized they had the opportunity to escape, he says, "They . . . stood to the southward until they crossed the line, where they met with westerly and northwesterly winds, which enabled them to sight [Samoa] without making a tack."

It sounds as if Dillon got his information about the *Globe*'s track from Gilbert Smith himself, and that Smith had favorable prevailing winds for much of his voyage. If he had, in fact, sighted some of the Navigators Islands, what would he have gained by stopping there? He was only beginning his journey, he had plenty of food and water, and he was on his way to summon help. From the Navigators at thirteen south latitude to Valparaiso at thirty-three south latitude was a straight shot in front of a northwesterly breeze.

On the map it seems so simple.

chapter twenty-six

The Mysterious Thain

Valparaiso to Nantucket, 1824

Given the horrors Smith and his men had endured and the tension of the unresolved situation lingering around them, it must have been sweet to be safe in the tidy port, which, according to a contemporary witness, had a "refreshing appearance, especially to mariners from a long voyage."[1] Almond and olive groves surrounded Consul Hogan's residence, and the city, composed mostly of whitewashed brick buildings with red tile roofs, stood handsomely against red earthen hills looming behind. However, it would be a while before Smith and his comrades enjoyed Valparaiso's comforts.

The consul had, as yet, no idea who

these men were or what their involvement was. As he was coming into port, Smith had confined the surly Thomas in irons, and so Consul Hogan found him. But what was he to make of this? Was Thomas guilty of mutiny and murder? Were there others as guilty or implicated in the crimes? What of the crew members they'd left behind? How many had been killed? By whom? Under what circumstances? Since there was no American naval ship at Valparaiso, Hogan sent the men of the *Globe* to an armed vessel anchored in the harbor, and locked them all in irons.[2]

This was the first exposure of the news of the mutiny to the world at large. The manner in which the information was elicited and presented would have much to do with the way the story was understood, and this, in turn, would determine what happened next. Fortunately for Smith and his comrades, Michael Hogan was a widely traveled man with a "warm cordiality of manners, united with . . . sprightliness and intelligence of his conversation, enlivened by anecdotes of all parts of the world." He was judicious and worldly, probably a career diplomat, certainly able to get to the bottom of whatever stories Gilbert Smith and his shipmates might put forth.[3]

Hogan allowed the crewmen a couple of days to get their wits about them, beginning his interrogations on June 9, 1824 — nearly six months after the mutiny. The questioning probably took place in his office on Valparaiso's single main street, in the commercial part of town close to the Custom House.[4] The young men of the *Globe* were exhausted, traumatized, and probably apprehensive of being associated in any way with the mutiny and murders. The consul, on his part, faced a doubly difficult task. Not only did he have to establish an accurate account of what had happened, he also needed to determine the criminal liability of each survivor, some of whom might lie to protect themselves. Against a backdrop of sparkling white buildings and fragrant gardens, each of the crewmen was subjected to an interrogation that would determine his fate.

Aside from establishing the basic facts of the mutiny, Consul Hogan's adept interrogations uncovered information that would have troubling repercussions back on Nantucket.

Testimony of the crew implicated Joseph Thomas in the mutiny. George Comstock, Peter Kidder, and Gilbert Smith all stated that Thomas had prior knowledge of

Samuel Comstock's plans. In his own deposition, Thomas admitted there was discontent on the Japan Grounds, but claimed he did not know its cause. He refused to blame Captain Worth for any of the troubles, stating, "The most the people had against him was his not allowing time enough to eat their victuals." He testified that his whipping at the hands of Captain Worth had taken place "3 or 4 days before" the murders.

Despite unanimous testimony from the other men that the beating had taken place on the day of the murders, Thomas attempted to distance himself from that event and from any anger he might have felt toward Captain Worth. Hogan blandly moved ahead, giving Thomas sufficient rope to see to his own hanging — if there was to be one.

The consul then assembled the depositions, along with his cover letter — stating that he thought Thomas was complicit in the mutiny, but that the other five men were innocent — sending copies to Secretary of State John Quincy Adams, the collectors of customs in Nantucket, the Navy Department, and Commodore Hull, head of America's Pacific Squadron, then based in Callao, Peru, about 1,000 miles north of Valparaiso.

While Consul Hogan had done his best to control the accuracy of the reports about the *Globe* affair, there was little anyone could do to keep rumors and informal accounts of the mutiny from spreading in the world at large. All of Valparaiso must have known about it, and word passed quickly through the naval and whaling fleets.[5]

On August 12, the *Globe*, which had been refurbished and remanned in the two months since her arrival, left port for America. The six survivors — Gilbert Smith, George Comstock, the Kidder brothers, Anthony Hansen, and Joseph Thomas — were aboard.

Despite Consul Hogan's best efforts, the rumors beat his official account back to America. On October 20, 1824, under the headline "Mutiny and Murder," the *Columbian Centinel* of Boston carried the first reports of the mutiny. The article mentioned Smith's escape, but missed Third Mate Nathaniel Fisher in the list of victims, and placed Humphries's hanging incorrectly in the sequence of events. Still, its overall accuracy suggests that the informant was close to the source, as indeed a sea captain would be expected to be privy to information circulating on the Valparaiso waterfront.

Variations of the article were reprinted up and down the eastern seaboard, appearing in such newspapers as *Boston Commercial Gazette* and the *New Bedford Mercury*.[6] The people of Nantucket and Martha's Vineyard could not have escaped being buffeted by the surge of fact, rumor, and misinformation surrounding the mutiny. Marine disasters were common enough; nearly every issue of every newspaper contained at least one article that began with the headline, "Tragic Loss . . ." But mutiny was uncommon, and the addition of mass murder made it a singular event in its day. The residents of both ports bore the stigma of being associated with such a repellent and fascinating occurrence. The only solace for the people of Nantucket was that all accounts cited Samuel B. Comstock's New York residence, rather than his birthplace.

A few weeks later, on November 14, the *Globe* arrived at Edgartown. Surprisingly, the first news from the ship was full of inaccuracies, as if the excitement, confusion, and horrific gravity of the event exerted a distorting force. The only survivors named by the *Nantucket Inquirer* were Smith and the Kidder brothers.

> Capt. Worth was killed by an axe, while asleep in his birth, by Thain, a sailor shipped at the Sandwich Islands. . . . Comstock, the elder, was hung at Mulgraves' Island. The younger brother was compelled to assist in his execution, and on his remonstrating afterwards, was beat to death with billets of wood.[7]

In the densely networked island communities on Martha's Vineyard and Nantucket, the sparse descriptions of axe murders and beating deaths must have caused agonies of grief and repugnance for the named victims' families. Worth, Beetle, and Lumbert horribly murdered. No word of Fisher. Mutineers and honest men (who was this Thain?) trapped together on a remote Pacific island. Almost all the early accounts state that eleven men (as opposed to the actual nine) were left on the Mulgraves, but no names are given. And what of the shame of the many Comstock relatives still on Nantucket, and their fellow Quakers, to whom all violence was abhorrent?[8]

On Nantucket in 1824, people had to deal with a reality that was unfolding faster and more horribly than they could have wished. Within days, the irrefutable testimony of the six eyewitnesses who returned on the *Globe* had begun to circulate. In its next issue on November 22, the editor of the weekly *Inquirer* published a retraction and correction that accurately followed the accounts of the survivors, down to the gruesome detail of the mortally wounded Lumbert hanging onto the edge of the deck and pleading for his life.

A week after the corrections, the paper ran another article including a reference to, and apology for, the misreporting of George Comstock's murder, "beat to death with billets of wood." That bit of misreporting must have occasioned a howl of outrage. The Comstocks were related to the Mitchells, and the Mitchells were players in the whaling industry. According to one Nantucket historian, the editor of the paper, "retired temporarily from the editorship with the last issue of 1824. . . ."[9]

By December, a more or less correct version of the story, drawn from Hogan's depositions and statements of the crew, had circulated throughout the East.[10]

These articles included the death of Fisher, omitted George's "murder," and, of course, printed every grizzly detail of John Lumbert's death.

On Monday, May 16, 1825, the spring session of the U.S. Circuit Court in Boston opened, Judge Joseph Story presiding. Justice Story was a Marblehead native with long experience in nautical matters. He bears the distinction of being singled out by Thomas Jefferson as the man responsible for the repeal of the Embargo Acts that were so damaging to New England's maritime interests prior to the War of 1812. In 1811, he was appointed an associate justice of the Supreme Court. At this time, Supreme Court judges also exercised circuit court jurisdictions, and Story's circuit included Maine, New Hampshire, Massachusetts, and Rhode Island.[11] It brought him many cases involving admiralty and prize law, and he soon became an acknowledged expert in this field. In 1841, he delivered the Supreme Court's majority decision in the case of the slave ship *Amistad*, ruling that the imprisoned slaves should be set free and allowed to return to Africa. He was exactly the right man to oversee the trial that would decide the guilt of Joseph Thomas, the lone crew member placed under arrest "on suspicion of being privy to the designs of the mutineers, previous to the attack."

However, in what must have been a stunning and bitter surprise to all of Nantucket and Martha's Vineyard, the case never moved past the grand jury. Judge Story, in his instructions to this body, pointed out that Thomas had not actually participated in the mutiny of which he stood accused. The other mutineers, the only witnesses who could provide evidence of his conspiracy to commit mutiny and murder, were either dead or on an island thousands of miles away and unable to be present. Given the circumstances, a trial would be futile. The grand jury deliberated for a few hours and agreed with the judge. No charges were pressed against Joseph Thomas.[12]

The article in the *Inquirer* for May 30 simply said:

We learn that THOMAS, the seaman charged with mutiny on board ship Globe of this port, after more than six months imprisonment in Boston, has been discharged — the Grand Jury not being able to find a bill.

Joseph Thomas walked out of that court-room and disappeared into the mists of history, but as far as Nantucket was concerned, the book was not yet closed.

Part Three

chapter twenty-seven

The Admiral

Washington, D.C., 2001

 One fascinating aspect of the *Globe* saga is the way it meanders and evolves; the tale can't be summarized in a single sentence. It was born in the flush of commercial excitement at the close of the War of 1812 and grew into a story of Nantucket's rise to dominance as a whaling port. Then it veered off wildly into the excesses of a homicidal maniac whose rampage was followed by one of the great journeys of survival at sea.

Powerful Nantucket oil interests involved the United States government, which, while seeking out the mutineers, navigated a complex political situation in South America and expanded America's sphere of influence in the Pacific. Under orders from Washington, a small force of intrepid navy men muscled their way through increasingly combative encounters with Pacific islanders, ultimately to discover the survivors of the *Globe* mutiny and accomplish a daring rescue.

Augustus Strong's manuscript journal of the *Dolphin*'s cruise — the one discovered in Vevay, Indiana — is the earliest account of America's rescue mission to Mili Atoll and of the fates of the mutineers and the innocent crewmen. Midshipman Strong recorded these events while they were taking place, and his journal has a raw energy that other, later narratives lack. However, there can be no telling Augustus Strong's story without first telling a little about the Admiral.

His name, actually, was George Emery, and he was a commander in the navy when I first met him. We booksellers didn't start calling him "the Admiral" until he made rear admiral in the late 1980s. I'm sure he was happy with his promotion, but we in the trade were happier, in our less complicated, greedy way. He'd make more money as admiral, and the nickname was an emblem of our optimism about where he'd spend it.

He was a collector with fire in his belly. He had the lust for acquisition that people in my trade find irresistible. But more, he had a scholar's passion for his material and an understanding of it that could range from deep philosophical insight to pure childlike delight. If he had a downside (and who doesn't?) it was that he was a Yankee, a State-of-Mainer, and was culturally — perhaps genetically by this time — imbued with the need to bargain. Many groups exhibit this tendency, each in its own fashion. Maine people tended to employ a wheedling attack, as if teetering at the edge of financial ruin, not particularly interested in the item at hand anyway, and married to the strictest of book collector's spouses, who probably wouldn't allow the prospective purchase in the house . . . so was that my final price? And would I consider taking half in cash and half in trade for these fine books (duplicates from his collection) that he'd brought for me to examine?

I'd met the Admiral's wife, and she was a lovely woman who approved of, and supported, with appropriate wifely detachment, her husband's collecting interests. And though I would always have preferred cash on the barrelhead, I was a Yankee too and rather enjoyed our negotiations. Truth be told, it was I who benefited in the long run from the goods I traded with the Admiral, as he knew full well. All he was really interested in was refining his collection — weeding out extraneous material and trading up for better copies of essential books.

The Admiral was where treasures like those from Irene's bunker ultimately landed. Although he bought on a smaller scale than, say, the institution to which Augustus Strong's journal went,[1] he was more focused. Each item in his collection bore an organic and meticulously documented relation to every other item. He did this because he loved doing it, but ultimately his concentration of rare materials, and the work of discovery he performed upon them, would go to an institution where it would be used by scholars and students of history, or it would go back on the market where it would entertain and inform the next generation. In an understated way, collectors like the Admiral conspired with pickers like Irene to single these artifacts out and assign them value. If they were valued, it was more likely that they would be preserved.

He was a genial, unflappable man with baby blue eyes, pink cheeks, and a gentle laugh. He had a relaxed way about him that made his military bearing seem the most natural thing in the world. The pressure of the handshake was perfect. The way he stood, one felt, was the way people *ought* to stand.

In his day job, he drove a submarine named the *Ohio*. The *Ohio* was armed with twenty-four ballistic missiles, each tipped with a hydrogen warhead, the two dozen of them representing more explosive power than had been expended in all of World War II — sufficient mega-tonnage to vaporize half the globe. A few years later, by then a vice admiral, he became Commander of the Submarine Force, U.S. Atlantic Fleet, a position that put him in control of all the Trident missiles on however many of those monster subs we had lurking around out there. I'd always been nervous about the idea of one man controlling such a devastating force. After getting to know the Admiral, I felt better about the whole situation.

The Admiral collected books and manuscripts pertaining to naval history, particularly the American navy's role in the War of 1812. He was deeply versed in his subject and often worked with the Naval Historical Center and the Navy Department Library in Washington. He'd authored a bibliography of historical manuscripts and had contributed to, and written the introduction for, a beautifully produced *Catalog of Early Imprints from the Navy Department Library*. I'd spoken with him about Augustus Strong's manuscript journal of the *Dolphin*'s rescue voyage, and he'd been most helpful in providing me with information about the navy of that era.

One day, shortly after the *Catalog of Early Imprints* was issued, I got an e-mail from the Admiral. He'd been leafing through it, he said, admiring its fine color reproductions, and there he came upon the entry for Augustus Strong's journal of his tour on the frigate *United States* and voyage on the schooner *Dolphin*. Now he could remember having handled it, discussing with the compilers whether it should be included in the book. He was embarrassed not to have connected it with the Strong manuscript I had told him about, so central to the story, but it had come back to him immediately when he saw its picture and description in the *Catalog*.

This implied that Strong's Vevay manuscript was not unique. I had sold it, for

316

quite a bit of money, as a one-of-a-kind item. If other copies existed, did that alter its value? How closely did the copy at the Navy Department Library correspond to the copy I had sold? And why, for crying out loud, were there two copies, anyway? I alternated between annoyance and fascination for a few days, and then stopped taking it personally. The only responsible thing to do was go down to Washington and take a look at that journal.

So, one exceedingly warm spring day, I visited the Washington Navy Yard in southeast D.C., and there, among the rows of handsomely restored 19th-century brick warehouses, found the library. The Admiral had told me, "Just use my name," and I found it worked quite well. The head librarian set me up at a worktable, then took me back into the vault. There, among the books, prints, documents, and manuscripts that recorded the key moments of America's naval history, was Augustus Strong's other journal.

On my way to Washington, I'd formed a theory that might account for the existence of a second copy of the journal. I knew that the *Dolphin* had gotten into a couple of scrapes on this voyage, and that there'd been an inquest at its conclusion. My hypothesis was that Strong had recorded events as they occurred in the first, Vevay, journal, then prepared an official account for the inquest, preserved in the second, Washington, journal. However, my theory was wrong.

Strong's Washington journal was written in two calf-bound, ledger-style books, with mechanically printed title pages that read, "Journal of a Cruise from Norfolk, Virginia to the Pacific Ocean, in the United States' Frigate UNITED STATES, Isaac Hull, esq'r, Commander." The blank calf-bound journals were standard items. The title pages and cover labels would have been custom printed by the stationer, who probably prepared similar books for every naval cruise that departed Norfolk.

Reading through the two volumes of Strong's Washington journal, I came to understand why they'd been fabricated in this manner. On the front blank of the second volume, in a hand quite unlike Strong's but very like that of John Percival, the ship's captain, was a paragraph instructing him on the proper way to keep a journal.

As part of their training in those days, midshipmen were required to maintain journals of their ships' activities. These were not personal or introspective documents. They were intended to take the form of deck logs — the standard way of recording a ship's position and daily evolutions. Though they make tedious reading, they provide the kind of information that would be basic to a military analysis of the vessel's activities. That was why such logs were kept, and why the proper keeping of them was a requisite skill for any young officer. And this was what I was looking at in Strong's Washington journal — his midshipman's practice version of the deck log of the schooner *Dolphin*.

In the Vevay journal, Strong recorded his personal narrative of events — the things he was experiencing, thinking, and feeling. In the Washington journal, he was keeping a strict account of procedures and events on board the vessel.

For example, on August 18, the day of the *Dolphin*'s departure, the Washington journal reads, "Commencing moderate breezes & cloudy Recd. From the American Ship Fame, a spar 27 ft. long & 6 in. in diameter. Hailed the Amer. Ship Alfred. Rcd. From Amer ship Providence ½ bbl of tar. At 4 P.M. got underway and stood out." The entry for the same day in the Vevay journal begins, "Standing down along the coast of Peru, with a fine breeze, the land in sight, distant about 15 miles. It has a very beautiful appearance."

Nor was this all. In his bibliography, the Admiral had cataloged five additional items written by Augustus Strong, but in fact there were nearly a dozen notebooks, fragments of letters, and other documents from Strong's navy days. Most of these were workbooks and compilations of notes on seamanship and navigation that Strong had used to prepare for his lieutenant's examination. However, a few were of a more personal nature. There was an unsigned note, obviously in a feminine hand, requesting the pleasure of Midshipman Strong's company, and there was a notebook that the Admiral had cataloged as "probably a work of fiction." This was a series of startlingly intense and direct letters regarding a duel between Strong and an antagonist named Bushrod Turner. It was rather disjointed, but I thought it had a romantic quality that went a long way toward explaining the subjective nature of the Vevay journal.

Finally, in a slim folder containing Strong's personnel records were carbon copies of two letters written in 1958 from the Navy Department historian to a Mr. T. A. Langstroth in Cincinnati, Ohio. The first told of the cruise of the *Dolphin*, and Augustus Strong's role in it. The second letter reiterated information contained in the *Navy Register* and Strong's old pay records: "He was born in Missouri and at the time of his appointment as Acting Midshipman, he was a citizen of and was appointed from the State of Ohio."

Cincinnati was just a stone's throw from Vevay, Indiana. It was a good bet Mr. Langstroth was at one time the owner of that Vevay journal.

chapter twenty-eight

Uncle Gorham

Nantucket, 1824

The abhorrence of mutiny by a maritime culture was profound. That such a thing should occur was bad enough; that news of it should be published was worse. That inaccurate news of it should be spread through the land was intolerable. This was certainly the feeling on Nantucket, and hardworking Gorham Coffin, old Christopher Mitchell's son-in-law, who saw the *Globe* through her earliest days, gave a voice to Nantucket's outrage. What made matters worse was that, in the cloud of rumor and innuendo surrounding the catastrophe, the actions of his nephew, Rowland Coffin, had been cast in a very unfavorable light.

★ ★ ★

In 1810, Gorham Coffin's brother Alfred died at sea. He left a widow, Peggy, and three children, one of whom was a four-year-old boy named Rowland. As was natural in that coherent society, some of the responsibility for the well-being of Alfred's widow and children devolved onto Gorham. The material extent of his support is unrecorded, but it must at least have included keeping a watchful eye on his nephew. He was certainly close enough to the family to be aware that the boy had come of age, because in 1822, when Rowland was seventeen, Uncle Gorham signed him on the *Globe* as assistant cooper.

This was an advantageous starting place for a traditional Nantucket whaling career. Young Rowland would learn a useful trade under cooper Cyrus Hussey, a Mitchell employee who was himself only seventeen years old. Hussey would be a friend as well as a mentor, and Captain Worth could be counted on to spare Rowland from some of the harsher aspects of life at sea. It seemed an ideal situation, one that surely won the approval of Gorham's sister-in-law, Peggy. They both must have been horrified to see it all turn to ashes two years later when Gilbert Smith, George Comstock, and Peter Kidder returned to Nantucket having pronounced Rowland a likely conspirator in the *Globe* mutiny.

During his deposition in Valparaiso before Consul Hogan, George Comstock mentioned "Rowland Coffin of Nantucket" who, he claimed, knew of the mutiny plot and acted as a spy for the mutineers. Peter Kidder corroborated the new information about young Rowland. He testified that "Rowland Coffin of Nantucket" had prior knowledge of the mutiny. Gilbert Smith also testified that Rowland Coffin knew of the mutiny, and intimated that he acted as an informant to the mutineers. "Coffin continued to be very much with the murderers after it had happened and carried to them all the information that passed forward."

The statements of Smith and his companions were made under oath before the American consul, and they could not be retracted. Still, Coffin tried to ensure that none of the speculations regarding his nephew were repeated or elaborated on in public. The temporary removal of the editor of Nantucket's newspaper may have been part of this strategy, and, judging by the fact that no public condemnations of

Rowland survive, he was successful in his efforts to stifle accusations against his nephew.[1]

Peter Kidder, the man George Comstock described as "very easily frightened," would have been susceptible to pressure from this august Mitchell partner. His talkative brother, Stephen, had said nothing about Rowland's complicity in his deposition. Better still, in a statement recorded by historian Obed Macy, Stephen said, shortly after his return, that all the crew, including Rowland Coffin, were "entirely unknowing to the intentions of any mutiny onboard of the ship."[2] Between them, Stephen Kidder and Gorham Coffin would have little difficulty gaining Peter's silence on the matter of Rowland Coffin.

In much the same way, Gilbert Smith could be convinced that there was no benefit in repeating his statement that "Rowland Coffin knew of it." He was widely regarded as the hero of the episode, the man who had engineered the escape. The Mitchell partners certainly saw him thus, and were indebted to him for rescuing the vessel in which they had such a tremendous capital investment.[3] Given his new standing in the company and in the community, it would be impolitic to make any further assertions that might deepen the scandal.

That left only George Comstock, the youngest, but probably the brightest and most articulate of the survivors. There was no advancement within the firm for him. Whatever the reasons — the shameful aura of scandal that now hung around the Comstock family, George's personal distaste for whaling life, or the trauma of his recent experiences — George Comstock did not return to the ranks of Mitchell & Co. employees. Significantly, George Comstock's unpublished manuscript account of the mutiny stopped short of accusing Rowland Coffin of complicity.

Probably during the same time that George was drafting this narrative (or dictating it to the Mitchell partners), Gorham Coffin was able to secure an affidavit from him that deflected attention away from Rowland and onto the motives of his accusers. In this statement, George held that, contrary to his testimony before Consul Hogan five months earlier, the fact of the matter was that "great jealousy was created against Rowland Coffin, from some of the crew, in consequence of his being taken the most notice of by the Captain." For the same reasons, he was "noticed" by Samuel Comstock and forced to do his bidding. George Comstock could thus accommodate Gorham Coffin without perjuring himself, and Coffin could use the affidavit to suggest that his nephew outperformed the others, and that they, in their jealousy, accused him of currying favor from Captain Worth.

This is the point at which the story moved from the homes of the four murdered officers, to all of America.

On November 29, a mere two weeks after the *Globe*'s return to Edgartown, Gorham Coffin sent a letter to Secretary of State John Quincy Adams. It was an impassioned, personal appeal "in vindication of the character of an injured orphan," his nephew, Rowland. "Having been left an orphan myself . . . I feel the duty more incumbent upon me, to assist in the protection of those in the same situation." In that both his parents were dead, Gorham Coffin was technically an orphan. However, his father, Abner, had died in 1802, when Gorham was twenty-eight years of age.

Uncle Gorham's letter sought to invalidate the sworn testimony of three witnesses by ascribing that testimony to petty personal motives. In its own time and context it may have sounded to Adams like an objective and sensible plea to avoid a rush to judgment. The advantage of hindsight reveals it to be more poignant, the arguments of a man struggling to preserve his family's honor. The fact that the U.S. secretary of state heeded it says much about whaling's place in American commerce.

Since the American Revolution, rapid growth of the young republic had been closely tied to expansion of her maritime economy. The whale fishery was a part of that expansion, with sperm and whale oils finding ever-increasing markets at home and abroad.[4] By the time Coffin wrote his letter, Nantucket had reached its zenith as the premier American whaling port. The annual catch had grown from 1,550 barrels of sperm oil in 1816 to 29,355 barrels in 1824.[5] It is not a stretch to say that Nantucket was the "Big Oil" of its day.

Gorham Coffin wasted little time following up his initial effort. On December 1, he wrote to Secretary of the Navy Samuel Southard. This letter repeated what he'd said to Adams, but added a request that the American navy take steps to rescue the crew and apprehend the mutineers. Gorham also enclosed in this letter a similar letter, which he asked Southard to forward to Commodore Hull, commander of America's Pacific Squadron.

Then, on December 22, he sent out his final letter. This one was addressed to Daniel Webster.

Webster was at that time beginning a stint as congressman representing the Boston district. He'd served in Congress some years earlier, then taken time off to pursue his career as a lawyer. He distinguished himself as an analyst and an orator in both arenas, and his persuasive abilities were already on their way to becoming a legendary part of American culture.[6] Hiring Webster then would have been like hiring Alan Dershowitz today.

Added to his abilities as a lawyer was Webster's standing as a politician. Uncle Gorham had now sought the advocacy of the State Department, the Navy Department, the Pacific Squadron, and the Congress of the United States, thus ensuring that every government official who received a copy of Consul Hogan's letter and depositions also received Coffin's rebuttal. Given the sensational nature of the crime and the manner in which news of it spread throughout the East, it would have been difficult to find anyone in Washington who wasn't aware of, and didn't have some opinion about, the mutiny on the *Globe*.

In his letter to Webster, Gorham Coffin repeated his claim that the testimony against Rowland was caused by jealousy. He attacked Smith and the rest of the crew: "Here we have six men, one of whom was an under officer, & not one of them with presence of mind, or courage enough to tie the two mutineers, put them in irons."

Then, in an exceptionally vituperative paragraph, he portrayed Smith as "so completely panick struck, that reason had taken its flight, & with it he flies, not knowing whither" — trying several hiding places aboard ship until he realized "that he had acted the fool long enough." Coffin says Smith then capitulated in a cowardly manner to Comstock. "I will do anything you tell me to, if you'll spare my life."

In closing, he apologized for having gone on too long, and feared he'd "been led by my feelings." To make it worth Webster's valuable time, he was enclosing a "trifle . . . not by any means considering it as a retaining fee; that will be more ample, should it ever become necessary."

What was it, then, if not a retainer? Earnest money in the best of interpretations; in the worst, a bribe. Of course, all this would all be moot until young Rowland came home.

On January 18, 1825, the citizens of Nantucket sent a petition to President James Monroe. It was signed by 142 of the island's leading merchants and ship owners, including the Mitchells. Significantly, Gorham Coffin was not among the signers. In fact, the petition had been circulating while Coffin was composing his letter to Webster, and he made a point of telling the congressman that he was not signing it, "believing it has not been address'd to the proper Department of the Government." This implies that Coffin was on his own in attacking Gilbert Smith and defending his nephew, but it also demonstrates that Nantucket as a whole was leaving no stone unturned in alerting Washington to the need for action.

The petition of the citizens was concise and direct. It summarized the events of the mutiny and the subsequent escape of Smith and his mates, pointing out how helpful it would have been if an American naval force were stationed closer to the scene of the crime.

During his last weeks in office, President Monroe responded by directing Secretary of the Navy Samuel Southard to take appropriate action on the matter. The secretary forwarded the memorial to Commodore Hull in the Pacific for his consideration. When, a few months later, newly elected President John Quincy Adams also received a memorial regarding America's commercial interests in the Pacific, Southard followed up by ordering Hull to send a vessel to the Mulgraves.[7]

By virtue of a well-executed and thorough campaign, Gorham Coffin and the oil industry of Nantucket made their needs known to the Monroe and Adams administrations. Their petitions for government assistance in the *Globe* matter dovetailed nicely with other requests from representatives of the merchant trade for a stronger American presence in the Pacific. Adams and Southard decided to utilize naval forces based in Peru to bring the mutineers to justice. While allowing America to render aid to her commercial people in the Pacific, this mission would also be a strong reminder of her power in those waters.

Then as now, when Big Oil talked, Washington listened.

chapter twenty-nine

Uncle Isaac and Mad Jack

Pacific Ocean, 1825

 In the early 1820s, Argentina, Chile, and Peru, led by such revolutionary heroes as José de San Martín, Bernardo O'Higgins, and Simón Bolívar, were struggling to free themselves from Spanish rule. During the first few years of that decade, it seemed possible that France and Spain would send a joint expeditionary force to crush the rebellions. This scenario alarmed the British, who suggested to the American minister in London that their two countries work together to keep France out of South America.

Jefferson, Madison, and Monroe seemed generally favorable toward this idea. However, John Quincy Adams, who was then

Monroe's secretary of state, thought that alliance with England in the United States' hemisphere would represent a tacit agreement that the Americas might be open to British intervention. Europe's business should be Europe's alone, and the Americas should be left to manage their own affairs. He presented his ideas so forcefully that Monroe ultimately embraced them, and even incorporated them in his annual presidential message to Congress, delivered in December 1823. His clear statement of Adams's argument became known as the Monroe Doctrine.

The situation in South America had been brewing for some time prior to this famous articulation of American foreign policy. In the years following the War of 1812, the Navy Department sent a succession of naval heroes — among them were James Biddle and John Downes — to Chile and Peru to oversee American interests. Given the volatile situation they faced and their own well-documented volatility as men of action, they did a remarkable job of protecting American commerce while negotiating the intricate politics of local struggles.[1]

Their job was made even more difficult by another character who had joined the dangerous mix. The brilliant, headstrong British naval hero Lord Thomas Cochrane had left England, and, in 1818, he took command of the Chilean Patriot Navy. While he added materially to their success, his bold and aggressive tactics repeatedly brought the Patriots to the brink of conflict with American naval forces. Until Cochrane's departure in 1823, news of his exploits regularly appeared in American journals and probably served as an inspiration to Samuel Comstock in his ill-fated attempt to join the Patriot Navy.[2]

In 1824, command of the Pacific Squadron was taken over by Isaac Hull. Known as "Uncle Isaac" by his adoring men, Hull had enormous responsibilities and slender means, beyond his intelligence and diplomatic ability, with which to accomplish them. Not only was he required to keep control of the situation in South America, the expansion of American commerce had increased the need for an official American presence in other parts of the Pacific. In particular, the Sandwich Islands were becoming a major stopping-place for American merchants and whalers.[3]

Hull's orders for his tour of duty as leader of the Pacific Squadron commanded him to "sail to the Sandwich Islands and remain for a few days, communicating with the agent of the United States there, and rendering to all our interests the best protection you can afford."[4] As for the rest of his mission, Southard told him, "This cruise will last three years, and it is intended that you visit the mouth of the Columbia River and return by the Cape of Good Hope." It was quite a tall order considering that the entire Pacific Squadron consisted only of the frigate *United States* with forty-four guns,[5] the sloop of war *Peacock* with eighteen guns, and the sloop of war *Dolphin* with twelve guns. Hull found the diplomatic situation sufficiently demanding that he never left South America.

A little more than a year after his arrival on that coast, in April 1825, he received a communication from the secretary of the navy expanding those orders. Not only was an official American presence required in Hawaii, but there was serious trouble about 2,000 miles to the southwest — on the Mulgrave Islands. He was to send a vessel there "as soon as the situation of the service will permit."

Commodore Hull still felt himself too deeply involved in revolutionary affairs to leave South America. Instead he sent his trusted lieutenant, "Mad Jack" Percival, in the armed schooner *Dolphin*, on a cruise unlike any ever attempted by an American naval vessel.

John Percival was a Cape Cod man, born in Barnstable in 1779. He was a mere thirteen years old when he first went to sea, and by the time he was twenty, he'd captained vessels in the West Indian and Atlantic trades, and been impressed into the British navy. In 1809, after sixteen years in the merchant trade, he joined the American navy as a "sailing master," a now obsolete rank roughly equivalent, in naval hierarchy, to lieutenant, junior grade, but having more influence.[6] Sailing masters were, as the title suggests, expected to be wise in the ways of the sea, and Percival certainly fit the bill. By this time he'd already begun to accumulate the strands of rumor and story that trailed him like a mane throughout his career. His exploits in the War of 1812 only added to their number.[7]

During the summer of 1813, New York was blockaded by a seventy-four-gun British ship-of-the-line. The warship's tender was particularly troublesome, harassing American fishing and commercial vessels at the southern end of the harbor. Percival borrowed a fishing smack from the locals, loaded her decks with produce and livestock, and hid thirty-two volunteers below decks. Off Sandy Hook, this target was stopped by the tender, only to be surprised by the sudden appearance of the Yankee musketeers firing away between sacks of potatoes and squealing pigs.[8] Largely on the strength of this escapade, he was able to transfer to the sloop-of-war *Peacock*, where he saw plenty of blue-water action, including the capture of a British warship.

Despite his many positive qualities, Percival was prone to feeling slighted. Perhaps because he had "come up through the hawse hole" rather than advancing in the company of gentleman officers, he was insecure about his origins. Edgy and hard-bitten, his temper made him an admirable fighter but frequently caused difficulty in dealing with his peers. Somewhere early in his career, he had picked up the nickname "Mad Jack." The fact that it stayed with him throughout his life suggests how little his experiences changed him. That he referred to *himself* as Mad Jack suggests how little interested he was in changing his behavior.

In short, he was a caricature of those larger-than-life officers who filled the ranks of the American navy during its heroic period.[9] In the years following the War of 1812, the ranks bulged with capable, aggressive officers, eager for glory and advancement, who had no chance of distinguishing themselves in battle. The annals of this era are rife with squabbles, insults, backbiting, and duels, and the swaggering officers that inhabit them rival Homer's heroes in their petulance.

Following a tour in South American waters, Percival was stationed at the Boston Navy Yard under Commodore Hull with whom, after a typically rocky start, he established a lifelong friendship. Then, from 1818 to 1820, he was again cruising those troubled South American waters. Apart from having his horse murdered beneath him by a Patriot soldier on shore,[10] Percival's tour of duty in South America seems to have been as routine as duty in that place would allow. The experience he acquired there, as well as his knowledge of Spanish language and customs, would stand him in good stead. When Hull received his orders to the Pacific Squadron, he had Percival transferred to his command as executive officer of the frigate *United States.*

By the summer of 1825, Commodore Hull was certain he'd be receiving official orders to send a ship to the Mulgrave and Sandwich Islands. His decision to send Percival and the *Dolphin* in his stead occasioned considerable maneuvering by officers and men for the chance to participate in such a potentially glamorous mission. In fact, Uncle Isaac himself had to use some subterfuge to remove Lieutenant Conner, the *Dolphin*'s perfectly capable commander, and have him replaced with Percival, the man of his choosing.[11]

As Hull made his choice of a trusted lieutenant so, it appears, did Percival. After transferring Lieutenant Conner, Hull had appointed two midshipmen as acting lieutenants on the *Dolphin*. Shortly thereafter, and undoubtedly with Percival's support, they were joined by Hiram Paulding, another officer from Hull's frigate *United States*. Paulding was senior to both midshipmen, and thus he became Percival's first lieutenant on the ship.

Like Captain Percival, Lieutenant Paulding was a man of considerable experience in South America. He'd served on the *Macedonian* during its tour as flagship of the Pacific Squadron in 1818, and accompanied Hull and the *United States* in 1824 for the current tour of duty in South America. Shortly after their arrival in Callao (the port city for Lima, Peru) Hull had selected him to carry dispatches to General Bolívar's headquarters in the Andes — 1,500 miles on horseback through South American wilds.

As well as being an adventurer and a sailor, Paulding was an author. In addition to his *Journal of a Cruise of the U.S. Schooner* Dolphin, he also wrote a book about his trek to visit Bolívar. Paulding was an intelligent observer, only mildly afflicted with the prejudices of his station in life. He seemed to have all the necessary qualifications for the *Dolphin*'s mission, and he did not disappoint.

Competition for the mission was even more intense because Hull determined that the crew should be reduced. The *Dolphin* was one of five clipper schooners[12] built in 1820 and 1821 to suppress piracy in the West Indies. They were sharp, relatively shallow of draft, and quite fast. At a little less than 200 tons and 88 feet in length, the *Dolphin* was considerably smaller than the *Globe*. However, because of her military function and the fact that she carried an enormous spread of sail, her normal complement of men was about four times the *Globe*'s crew of twenty-one men.[13] This crew was shoehorned with military efficiency into the schooner, but military standards of diet and cleanliness ensured that they would live at least as well as whalemen — especially since they weren't competing for space with tons of oily blubber or hundreds of casks.

As far as this mission was concerned, the ship's configuration and complement of men presented another problem. A large crew on a small vessel meant frequent stops to resupply. This was easily accomplished in tactical situations such as chasing pirates in the Caribbean, but it presented serious difficulties when cruising across thousands of miles of empty Pacific.

Early in August, Hull ordered Lieutenant Percival to provide him with an estimate of the minimum number of men it would take to man the ship and operate her twelve six-pounder guns.[14] Percival determined that sixty-seven men would be the minimum crew required to sail and fight the *Dolphin*. It was probably no coincidence that the provisions this crew required for six months matched the ship's storage capacity nearly exactly.[15]

On August 14, Percival got his orders from Commodore Hull. Although expressed in a breathless, rather mangled syntax, their intent was clear enough. "Use all the means in your power to ascertain whether the mutineers of the Globe are still on the island and should it appear that they are you will use such measures as you may appear best calculated to get them on board your vessel and to secure them preferring a mild and friendly course as regards the natives of the island to that of using force to attain them."[16] He went on to suggest that if the mutineers refused to surrender, Percival might kidnap a few of the native chiefs and hold them until the natives offer their assistance in rounding up the fugitives.

The mutineers, Hull said, had two whaleboats and might have left the island. If that were the case, Percival was to find and apprehend them. Then, if he had sufficient provisions, he was to stop at the Sandwich Islands and determine the state of things there, particularly as regarded trade privileges, "and as the natives owe large sums of money, it would be desirable to ascertain whether the present king or individuals of the islds. that owe American citizens have the means if they have the disposition to pay their claims."

Sailing to paradise to capture mutineers! It sounds like a grand adventure, and it was considered so by the crew, as competition for sites on the ship would attest. But it was also regarded as a perilous cruise, the outcome of which was uncertain. Augustus Strong wrote to his brother, "Several imparted to me that it was impossible we should ever return, and in fact the Commodore expressed his doubt of the same — These reflections I did not allow to disturb me until the dangers were past, and then reflected upon them with a kind of pride that we had risen superior to them."[17] Officers may have sought duty on this hazardous mission as a means of advancement rather than an opportunity for adventure.

This was not the modern navy, with its highly organized systems of support and supply. Provisioning the ship was an uncertain process involving bartering with locals along the South American coast. Adequacy of food and water was a constant anxiety. At the climax of their voyage, the crew of the *Dolphin* would find themselves in an alien, uncharted environment faced with the task of subduing desperados who had

brutally murdered six men, and who might have recruited savage tribes to their cause.

Their itinerary had them stopping at the Galapagos Islands, off the coast of Peru, then sailing down to the Marquesas to resupply, northwest up to the Mulgrave Islands (Mili Atoll) to capture the mutineers, over to the Sandwich Islands to assist in America's commercial relations, then home to South America. First, however, they needed to pick up supplies with which to begin their voyage. On August 18, 1825, the *Dolphin* sailed from Chorrillos, on the coast of Peru. Next day she stopped at the harbor at Casma; then sailed a short distance north to Santa where she obtained supplies; then on to Huanchaco and Paita, where she took on more.

Finally, on the second of September Mad Jack and his crew headed west into the Pacific.

chapter thirty

Augustus and Herman

Marquesas Islands, 1825

 Augustus Strong had literary ambitions. Laid into the pages of his Washington journal was the single sheet of another composition, a fanciful and allegorical voyage in the ship *Invincible* "to explore the harbors, rivers, shoals & islands along the coast of the land of Matrimony." The ship was manned by "the intellects of 21 persons, who were qualified to trace out every danger to which man is subject and with which this coast abounds." Their first port, in Strong's story, was the island of Bashfulness, where they were coldly received by the female inhabitants. Mercifully, the story broke off at that point, a little short of filling the page.

346

There was also Strong's story of the duel, which the Admiral had cataloged as, "Notes passed between Strong and Midshipman Bushrod Turner, 1828. Regarding a proposed duel (probably a work of fiction)."[1] Finally, there was the internal evidence of the Vevay journal itself. On several occasions, Strong rewrote picturesque or exciting incidents. Invariably, the second version was more highly colored, decorated with tropes and philosophical asides.[2]

This is not to belittle his efforts. Born in Missouri and raised in Ohio, Augustus Strong had a head full of the Midwest when he showed up in Norfolk as a green midshipman in 1823. The cruise of the *United States* around the Horn must have been an eye-opener for him. Once he was selected for the *Dolphin*'s mission, he was bound for the adventure of his life. His desire to record that adventure is an appealingly human one, and his literary aspirations ensured that he'd put sincere effort into it.

After a sail of four days from the South American coast, the *Dolphin* made Hood's Island in the Galapagos, and Strong's adventure began in earnest. These were the *Islas Encantadas* of the old Spaniards, the "enchanted islands" described by Herman Melville in his heavily atmospheric tour de force, *The Encantadas*. The unearthly parched and blasted aspect of this volcanic archipelago inspired awe then and still does; it is probably the only site in the whole saga of the *Globe* mutiny that remains to this day much the same as it did in 1825.

Hood, or Española, Island is the most southeasterly of the group. Unlike some of the larger islands, it is low and flat and has as its most distinguishing feature the spectacular, stark, sandy beach of Gardner's Bay. Here Mad Jack and his crew anchored on September 6. Though inhospitable, this island was much frequented by whalers because it abounded in tortoises.[3] These animals could survive for long periods with no food or water, and because they were edible, they were captured in great numbers and taken aboard ships as food. Percival sent a party ashore to capture a supply.

"Immediately on our anchoring," Strong says, "we sent a party of men on shore to procure Terapins." Strong and Lieutenant Homer[4] wandered off by themselves, "having a desire to see everything that was curious or remarkable on the Island." They killed a snake and saw hundreds of lizards. There were doves, small and beautifully colored, and another species reminiscent of the blue jay. "I have sat down upon a stone and these birds have come up and eat out of my hand. In fact all the birds of this Island, are so very tame, that a man possessed of any humanity could not have the heart to kill them." And more wonders: "The stones are so burnt, that you may strike two of them together, and they will retain the sound as long as bell mettle."

As for the "Terapins," they lived in a valley about two and a half miles inland from the anchorage. In this valley grew large cacti — Strong called them "trees," — that provided sustenance for the tortoises. "Several of them join together and in the course of a few weeks or months they knaw the tree down, which is about 4′ in circumference." As plentiful and easy to catch as this quarry was, "our men suffered much in carrying them. They make them fast by a fore and a hind leg and sling them on their backs as a soldier does his knapsack." The hunting party turned into something of a forced march as the crew lugged the 100-pound tortoises over two miles of cinders to the ship. "We got on board upward of an hundred and fifty, which lasted us until we arrived at the Washington Groupe [the Marquesas]."

On September 10, they sailed about forty miles westward to Charles Island (what is now Floreana). This was the site of the famous Galapagos "Post Office" established by British and American whalers and used with great success by Captain Porter to gain intelligence about the movements of the British whaling fleet during his cruise in the War of 1812.[5] Strong remarks, "We sent an officer to the post office, and found several old letters, which had been left there by whalers."

Next morning, with 150 tortoises lumbering about on deck,[6] they got underway, headed southwest, with a "fine breeze" behind them.

They made the Marquesas on September 26, landing at Hiva Oa, known in those times as Dominica. Mad Jack and Lieutenant Paulding took the gig into the harbor and did a little trading for fruit, after which the *Dolphin* resumed her journey, passing the smaller Fatu Huku to the north, then heading west.

By the next afternoon, they had reached

Nooaheeva (Nuku Hiva), the classic Marquesan island, whose misty heights and lush valleys afforded a perfect contrast to the Spartan Galapagos. They put in at a large harbor, which, in 1792, had been named Comptroller Bay by an indefatigably Anglo-centric British naval officer.[7] Strong reported, "The passage is narrow and deep. On both sides are lofty hills . . . from the entrance of the bay is visible a high point which projects some distance out, and forms two beautiful vallies, and bays, owned by two different tribes." As attractive as they might have seemed, these highlands caused capricious winds. It took the *Dolphin* upwards of two hours to work the mile and a half into the southernmost bay, "continually wearing, tacking, boxing and bracing."

As soon as they entered the bay, outrigger canoes approached. Strong had apparently never seen such craft and was unaware of their excellent seagoing qualities, since he describes them in detail and comments on the fact that they were "rudely constructed . . . in pieces, sewed together, not so neat, but that it requires a man almost constantly bailing." Five young men pulled up next to the ship in one of the canoes and sang out, "Typee, Typee"

351

while pointing to the north.

Shortly after another canoe came near us with the same number of men in it crying Happa Happa, and pointing to the southward & westward — The Captain got up on the stern and pointed toward Happa Bay, repeating the words at the same time. The one in the bow of the Typee canoe immediately raised himself up, extended his hand toward the Happas', repeating in a very melancholy tone, Happa, Happa.

The Haapa and the Taipi tribes were mortal enemies. Each was trying to dissuade Percival from landing at the other's beach. The description of the bay and the *Dolphin*'s encounter is fascinating because it prefigures an encounter described seventeen years later by another young sailor. His name was Herman Melville, and his whaleship was visiting "Nukuheva" for the first time. "Immediately adjacent to Nukuheva," Melville wrote, "lies the lovely valley of Happar, whose inhabitants cherish the most friendly relations with the inhabitants of Nukuheva. On the other side of Happar, and closely adjoining it, is the magnificent valley of the dreaded

Typees, the unappeasable enemies of both these tribes."[8]

Melville jumped ship and lived among the Taipi for nearly a month (though in his narrative he stretched it to four). Strong and his shipmates only spent two weeks in the Marquesas. Yet, so powerful was the effect of this place, and so little had it changed in the time between their visits, that the images evoked by each writer resemble one another, and their descriptions run parallel on many particulars. For example Strong says, "The manner in which they are tattooed is certainly very remarkable . . . executed with great taste, judgment and ingenuity; but remarkable as it may be [it] bears no resemblance to any thing that can be mentioned." Melville observes, "But that which was most remarkable in the appearance of the splendid islander was the elaborated tattooing. . . . All imaginable lines and curves and figures were delineated over his whole body." Where Melville goes on to describe these tattoos, Strong, in his more pedestrian but no less entertaining manner, provides a description of how they were actually applied. Melville gives a sensuous description of his female lead, Fayaway. Strong, for his part, has as much to say about the tattooed Marquesan women, "their feet, and as high as the calf of their leg, their hands and shoulders are beautiful past description. . . ." Both mention Captain Porter, America's naval hero in the War of 1812. Strong's "Happas" revered him. Melville's "Typees" fought him.

As powerful and exotic as these sights and scenes must have been, as novel as the landscape seemed, as strange and new as the smells and sounds were, still there was no occasion more powerful than the actual physical contact of the two races. In this most intense and intimate moment of meeting, Strong notes their curious behavior:

> They have frequently come up to me and felt of my clothes, separated my shirt, looked attentively at one another for a moment, and then started back in the greatest surprise.

Melville elucidates:

> Their surprise mounted the highest when we began to remove our uncomfortable garments, which were saturated with rain. They scanned the whiteness of our limbs, and seemed utterly unable

to account for the contrast they presented to the swarthy hue of our faces, embrowned from a six months' exposure to the scorching sun of the Line.[9]

Strong was never a source for Melville, yet there is something reassuring about having Midshipman Strong's observations reinforced by one of America's great authors.

After a few days, the *Dolphin* sailed the seven miles from Comptroller Bay around to Massachusetts Bay.[10] Strong describes in detail the houses in the villages there, the funeral customs, the concoction and intoxicating effect of kava, and the swarms of curious and friendly little boys that accompanied them everywhere. On the fifth of October,[11] Strong and Lieutenant Paulding sailed a few more miles down the coast in one of the ship's boats to Lewis Bay. Paulding led the mission, which was ultimately unsuccessful, to get hogs for the ship, and writes an exciting account of their difficult row back to the *Dolphin*. Strong, on the other hand, slips into a reverie about a farm that he sees on shore.

The ship departed the next day. However, after noting this, Strong's journal reverts to September 28 and the landing at the Marquesas. He wants to take another crack at describing the remarkable place; this time his descriptions are compressed and rhapsodic. "The scenery here is more sublimely romantic than even the most fruitful imagination can describe. . . ." A big part of this sublime romance were the Marquesan women, whom he pictures continually coming on board the vessel, "with no other clothing than a few leaves attached together with grass. . . . The females are generally very handsome; but are so perfectly in a state of nature, as to have no ideas of modesty or feminine delicacy." The women must have utterly fascinated Percival's men. The conventions of Strong's day inhibited him from saying too much about their lives there, but one gets the sense that their tour of the Marquesas was charmed and easy.

In his second, more literary description of their arrival in Massachusetts Bay, Strong adds a part of the story he'd neglected his first time through. Among the natives, the *Dolphin*'s crew discovered the two beachcombers "in the garb of sailors — When these men came on board they informed us they had been here some time, having left an English Whale Ship. They understood the language very well had attached to the manners and customs of the people and by no means wished to leave the place." Two white men were living a benign version of Samuel Comstock's dream.

If there was a dark side to life in the Marquesas (other than such "filthy practices" as eating the lice they groomed from one another), it was the perception of cannibalism. Strong spends a full page in his rewrite dealing with the subject. "It is well ascertained that these people eat the bodies of their enemies taken in battle." The women, he says, do not partake of human flesh, but this is by choice rather than by taboo. As for the dietary habits of the men, he cannot say whether this flesh eating is done "to gratify a feeling of revenge, or whether . . . they eat it in preference to other food." However, he does have the irrefutable report of an informant (perhaps one of the English whalemen) who was offered the rib of a "roasted human king." This man's hosts "spoke of it to him as a delicious food, and appeared to eat of it as a feast."

Of their departure from the Marquesas, Strong wrote movingly about the attachment he'd formed for the natives, then launched into an apostrophe that landed well short of Melvillean heights.

Indeed one who had ineffectually struggled in the storm of adversity — whose hoary locks and spirit shaken by disappointment had proved the — [Strong cannot find the right word here, so he leaves a blank space] — of his fortitude and perseverance might here find assylum from the reverse of fortune, from the — [a similar blank] — luxery and the pomp of society; here he might terminate his days in the tranquility of retirement; having gained the affection and reverence of those children of nature, without being interrupted by the broils of civil discord — Though bereft of gorgeously sculpted monuments erected to his memory by the sycophant; yet free from the machinations that grand god, monster jealousy its inseparable associate slander, and the kindred relatives Tyranny, and oppression —

The passage stops there — the maundering of a young man who'd found a spot of innocence and repose in the hard world of floggings, mutiny, and murder. It sounds as if he might have fallen in love on

358

Nuku Hiva; certainly, the place spoke to his soul. As purple as that final paragraph might be, it doesn't reveal a small or spiteful nature. In fact, it makes one rather fond, in an avuncular way, of Midshipman Strong.

chapter thirty-one

The Ambassadors

Tokelau Islands, 1825

After departing Nuku Hiva on October 5, the *Dolphin* sailed for a week through squalls and generally rotten weather, in Strong's words "changing our course almost every hour, probably owing to an expectation on the part of the Commander, of making a discovery of unknown land, which is supposed by some navigators to exist about this latitude." Percival was not seeking the glory of new discoveries in these waters. He was performing a tedious but necessary task, which would ultimately result in safer navigation. Whalers and merchantmen frequently reported sightings of islands and hazardous reefs, but were ill equipped or too preoccupied to accurately chart them. Ascertaining that an island or reef did not exist in a particular location was just as important as discovering that it did.

Finally, on the eleventh, they sighted Caroline Island, an uninhabited atoll situated between the Line Islands and Tahiti. While Strong was enthralled with the richness of aquatic life in the coral reefs, he noted that there was no good source of fresh water on the atoll. This underscored the *Dolphin*'s greatest vulnerability as she sailed the expanse of the Pacific. She could only carry about 5,000 gallons of drinking water, and her crew expended as much as a hundred gallons daily. The one statistic recorded even more closely than their position was the daily consumption of water.[1]

They continued west through more bad weather, and spent several days in an unsuccessful search for another island until they caught a fine westerly breeze that brought them to Nukunono, or Duke of Clarence Island, on the twenty-ninth of October. Midmost of the three islands in the Tokelau group, north of Samoa, Nukunono had been little visited by traders or whalers, and its inhabitants presented a startling contrast to the more decorous Marquesans.[2]

Shortly after the *Dolphin*'s arrival, they were surrounded by dozens of canoes, some of whose passengers climbed aboard the ship. One fellow boldly detached an iron fitting from the ship's boat and was about to make off with it when Mad Jack snatched it from his hand. To his immense surprise, the native grabbed it back and jumped over the side with it. Strong wrote, "The Captain immediately ran to the stern and discharged a piece at him, but this did not appear to alarm him." Meanwhile, more mayhem was breaking out. Another native seized the log reel and tossed it down to his canoe below. An officer grabbed him by the hair and Lieutenant Paulding ran over and gave him a whack with a fowling piece. The native responded by trying to take the gun from him. When Paulding summoned help, his antagonist escaped into the water. Percival ordered a volley of muskets and then the "great guns." But so innocent were these people of firearms that they scarcely noticed the gunfire amid the general excitement, and it had little effect in dispersing the throngs of canoes. The natives threw coconuts at the ship (one of them bonked the surgeon on *his* coconut), and the crew returned these missiles, more with annoyance than real hostility.

All this while, the *Dolphin* had been running along the coast looking for an anchorage. When she'd try to sound for depth, the natives would cut off the sounding lead and steal the line attached to it. After about an hour, Percival found a likely spot to anchor the ship, and this time sent out a boat to take the soundings. As soon as the boat's crew commenced their work, they found themselves completely surrounded by canoes. When they threw out their sounding lead, the natives once again seized it. Realizing this was a threatening situation, the *Dolphin*'s boat attempted a retreat. But the islanders' canoes blocked the way. Some of the natives seized her oars. Lieutenant Homer drew his pistol, and one of the midshipmen (Strong does not identify him) "fired & shot the man . . . through his hand; he immediately jumped from the canoe into the water, holding his bleeding hand to the view of his companions."

Their newfound awareness of the true function of firearms was sufficient to keep the islanders at bay, though a large crowd of both sexes waited eagerly on shore. If the boat had landed, Strong remarked dryly, "They would undoubtedly have had a fine roast ere morning."

The ship managed to anchor, and by afternoon the canoes returned, this time to trade, "assuming such conduct as convinced us they were sensible of our superiority." They even saw the shooting victim, "not seeming much concerned about the wound, yet appearing to be conscious of guilt." Percival made him a present of some pieces of iron, "but nothing could have induced him to come on board, where he might have received the assistance of the surgeon." The rest of the islanders, owing to their "roguish inclinations" proved to be difficult trading partners, as they continued trying to steal anything not nailed down.

Strong's descriptions of this and future encounters abound in misunderstandings and faulty assumptions on the part of navy and native alike. The two sides played by such vastly different rules that the situation was continually adrift. Futile gestures flew like wild shots from both sides, and matters veered from comic absurdity to mortal danger and back. The only constant was that if the encounter lurched close enough to personal hazard, the guns came out.

Percival's men were not killers. They were aware that their actions might have consequences for future visitors. But they were not anthropologists. They were military personnel on a mission that had as its ultimate goal the enhancement of America's political and commercial interests. They were a small part of what Melville referred to as the "all-grasping western world,"[3] and they had little patience for individuals or societies who did not share their worldview. Still, even the white men learned. Having studied Marquesan canoes, Strong provided his prospective readers with a much more knowledgeable and sympathetic description of the outrigger canoes he saw on Duke of Clarence Island.

The next day they sailed to Duke of York Island, northernmost of the Tokelau group, and traded for supplies and curios with the timid natives there. They found a few metal artifacts ashore, suggesting that white men had visited the island, "though this place is very little known by navigators." Then, continuing west through more rough weather, they made Byron's Island on the ninth of November.

This island was named after its discoverer, John Byron, the poet's grandfather,[4] who first landed there in 1765. (One of his ships was also named the *Dolphin*.)

Percival and his men found Byron's Island reminiscent of Duke of Clarence Island, and so were its inhabitants, swarming around them in outrigger canoes. Once aboard, it was difficult to get them off again, and they exhibited the same predilection for thievery. The women were very handsome, and the men were armed with spears and knives edged with shark's teeth, "extremely sharp . . . the teeth resembling a saw." On the morning of the tenth, the captain decided to go ashore in the ship's gig to search for a watering spot. Strong recalled, "When the gig came near the shore it was surrounded and assaulted with stones, one of which struck a man in the head and knocked him down — some even threw their destructive spears at the boats crew . . . it was found necessary to fire among them." The natives remaining aboard ship had to be forcefully driven off, and the *Dolphin*'s guns were run out, in case a broadside of grapeshot should be called for. But Percival's gig made it back safely, and the anchors were hauled up.

Just as they were about to sail away from this unwelcoming place, one of the islanders crept aboard and, while all hands were occupied getting underway, stole a musket. Despite a barrage of gunfire that wounded several innocent natives, the culprit made it back to shore and ran up and down the beach waving his prize triumphantly. It had been a bad morning for Captain Percival. He fired a broadside into the village, then detailed three boats full of armed sailors to recover the musket.

Off they rowed, a fired-up Mad Jack in the lead. His boat landed, discharged him and his avenging crew, and immediately got swamped in the surf and stove on the coral rocks. The other two boats hovered nervously offshore while Percival and his party collected themselves and considered what to do next, since some of their muskets had gotten wet and they were surrounded by a throng of angry natives. As Strong says, "The situation of those on shore was now rather critical."

However, Mad Jack had been in tight scrapes before. He drew his men up in a defensive line and attempted to negotiate while they got their muskets back in working order. His efforts unheeded by the menacing natives, Percival called in fire from the ship. "A continual firing was kept up until after meridian [noon]. The natives probably alarmed at our perseverance brought the musket back." Unfortunately, it was missing its lock and bayonet. Mad

Jack then directed his gunners to fire away at a large hut "where multitudes of the Natives had assembled, probably for the purpose of giving battle." Trees began falling, and the natives scattered. Mad Jack advanced inshore, both to show the natives that he was unafraid, and to search for water. (He found none.) One of the men was hit by a stone and severely injured. His attacker was shot. By late afternoon, the tide was high enough that a boat could be brought ashore. After nearly capsizing in the surf, the weary party was evacuated.

The men who'd spent the day ashore apparently gave Midshipman Strong new information, and he duly rewrote the account in his journal in a more organized and inclusive manner, even giving it a dateline — "Bryon's Island. Nov. 9th." It was an exciting narrative, and it would have made an interesting piece in some contemporary newspaper or journal. Apparently Strong was pleased with this effort, because he signed it at the bottom in a fancy, formal hand, "Midshipman A. R. Strong."

The *Dolphin* then sailed west for a day to Drummonds Island, and again they got into difficulty with the natives. One of them snatched a hat from a sailor's head and swam away with it, under the by now familiar hail of musket fire and volleys from the great guns. Another islander stole a piece from one of the guns and made his getaway. By this time, Mad Jack had reached his limit. While gunfire from the ship pinned the thief down in his canoe, Captain Percival went over in a boat and captured him. He brought the man back to the ship and whipped him. "Four dozen was inflicted with a cat of nine tails, on his bare back; but he did not appear to mind it much, although he had a charge of shot in his back, until the flagellation was nearly over — then he look[ed] up to the Boatswain's mate in a most supplicating manner — without uttering a syllable." The severity of the beating might have killed a white man. They sent him over the side again and sailed away.

It had been a brutal and unhappy series of encounters for the men of the *Dolphin*, but it prepared them for the worst they might expect at their next stop, the Mulgrave Islands.

chapter thirty-two

Midshipman S.

Mili Atoll, November 23, 1825

On November 23, the *Dolphin* paused near the middle of the long strip of island and sent a cutter in. When the boat reached the first line of breakers, the search party clambered out and waded to a white coral beach. The equatorial sun hammered down on their backs as they slogged ashore. The water did not cool them. In this place nothing was cool.

They stopped above the tide line to assemble their gear — cutlasses, pistols, ammunition, water, a day's rations — and soon felt the heat of the sand working up through their boots. Two of the men then headed north.[1] The other two, Sailing Master Arthur Lewis and Midshipman

Augustus Strong, began walking south, the land pitching under their feet as it does for men who have been long aboard ship.

The island, one of many in the atoll, was ten miles in length and half a mile wide. Down its center ran a strip of jungle — a tangle of soaring trees competing for sunlight with greedy vines and shrubs. Eventually Lewis and Strong discovered a place where the vegetation thinned sufficiently to let them enter. Here they were protected from the sun, but the air was thick, difficult to breathe, and the flies found them. Water from the island's opposite shore glinted through the trees. It was luminous and impossibly green.

In America there was no such color of water, nor such woods. Parasitic ferns nested in clumps of humus rotted into the massive trunks of the breadfruit trees. Leaves like devil's ears wagged in their faces. Sinewy black arms of roots bulged from the coarse gray soil. Every few steps the undergrowth rustled and twitched as land crabs scuttled to safety. Bold rats with tufted fur tails, smaller than ship's rats and nimble as roaches, scooted everywhere on frenzied errands. It was all overwhelmingly new and difficult and strange, sufficient cause for a man to move carefully. But

Lewis and Strong had a better reason for caution.

Somewhere out there, the mutineers were lying in wait. How heavily they were armed and how many native accomplices they had, God only knew.

Four days earlier, on the evening of November 19, 1825, the *Dolphin* had made her first landfall, probably off Chirubon, at the southeastern end of Mili Atoll.

Southernmost in the 700-mile-long chain of the Marshall Islands,[2] Mili is a ring of about ninety islets of varying sizes, strung like elongated beads in a necklace around a central lagoon. A few channels connect the vast central lagoon with the ocean. The islets themselves are slender strands, many only a few hundred yards across, the greatest barely a mile wide. Most are covered with shrubby undergrowth interspersed with groves of coconut, and breadfruit and pandanus plants.

The roughly circular configuration of the islets is due to the fact that they were built upon the rim of a long-extinct volcano. Over the eons, coral had grown on the volcanic rock and eventually risen above sea level. The actions of weather and waves broke the coral down into sand, and by whatever miracle such things occurred, the life-sustaining flora of the island had found its way there, to be followed much later by birds, men, and rats.

Mad Jack Percival and his crew had sailed halfway across the Pacific, been tempted by the seductive Marquesas, and survived hostile encounters with inhabitants in the Tokelau Islands in order to get to this place. Now, as Midshipman Augustus Strong wrote in his journal, their "great and chief object . . . was to commence the search for the mutineers of the American Whale Ship Globe . . . without loss of time."

To the men in Washington who issued these orders, Mili Atoll (known then as the Mulgrave Islands) seemed no more than a flyspeck on a chart — the abstract objective of a straightforward mission. To Mad Jack and his men, the experience of the place was quite a different matter. They didn't know they'd landed at Chirubon, because no one had yet charted the atoll. They were unaware, at first, that it *was* an atoll.

This much they knew: it was vast. Rich blue Pacific waters broke on a coral shelf about one hundred yards in width, beyond which was beach, then jungle. The white-and-green hump of land stretched west-ward[3] as far as they could see, until it bumped the horizon and disappeared in the mist of the surf. In the distance to the north, unconnected islands seemed to cluster. In fact, they made up the opposite rim of the atoll, but to the early explorers, they seemed more like a "range" or "groupe."

Over the next two days, while they replenished their precious supply of fresh water, set their ship in order, and made their first tentative explorations, they came to realize that this place was similar, in its general contours, to others they had visited.

Although the topography was similar, Percival's men soon discovered that the people were different. The natives they saw in their first watering trips ashore were not as tattooed as the inhabitants of other islands. They were of medium height and blackness, with long hair, which they piled atop their heads. They wore necklaces, grass loincloths, and large plugs of rolled leaves inserted in holes in their ear lobes. Furthermore, Strong reports, "They evinced a kinder and more honest disposition than any other of the Isls we had before visited — They also showed a considerable degree of timidity."

Initially, this "timidity" of the islanders was striking. In trade, they'd yield up their finest goods for any bit of metal or cloth. But they attempted to hide their most precious ornaments, and seemed to be trading only out of fear. The crew finally managed to coax a few natives on board, and eventually their shyness wore off. In this new mode, they revealed themselves as very tough traders indeed. Strong says they wouldn't part with anything "unless receiving double its value in return; so that we found they were not unacquainted with bartering as we at first supposed."

On the morning of the twenty-third, Percival sent the first search parties ashore.[4] Second Lieutenant Homer and Purser Bates headed north, and Midshipman Strong and Sailing Master Lewis went south. They found no sign of the mutineers, and the tension of the strange landscape was soon relieved by the hospitality of the natives, one of whom invited Lewis and Strong into his hut. His wife entertained them, "playing upon a drum and accompanying it with her voice." When they left, some of the women showed them the path, much to Strong's delight. At every hut they passed, they were given coconut milk and invited to enter.

Following their return that afternoon, the mood began to change. Another party, closer to the anchorage, entered a hut and found "an old whale lance, and a piece of duck, part of a pair of trowsers." Lieutenant Paulding commented, "Whenever the subject of our visit was pressed upon them, by pointing to the whaler's lance, they became silent, pretending to be ignorant of our meaning."

Having searched the vicinity of their first anchorage, they sailed slowly west, along the southern string of islets, occasionally stopping and exploring. Mad Jack went ashore and found a few exceedingly wary people. At one point they saw what appeared to be a flag flying from the top of a tree. Strong reports, "All joyfully exclaimed, 'it's the American flag; there they are!'" But they were mistaken. The flag was actually a native's mat.

On the afternoon of November 24, they anchored at Lukunor, one of the major passages into the inland sea. On the beach at the edge of this inlet, says Strong, "were a dozen or more large canoes hauled up."

These were impressive, seaworthy craft, with "sails as large as those of the largest sailboats in our country." The captain went ashore here, and on examining some of the canoes, he found that parts of their rigging were fastened with pieces of cloth, and that some of their masts had been made from the *Globe*'s spars. Many natives had assembled here, but according to Strong they seemed uneasy about Percival's examination of their canoes. At dawn the next morning, the canoes sailed away in a flock, as if they were fleeing.

Seeing this, Captain Percival realized that as long as the natives were able to crisscross the lagoon, they would be able to evade his landing parties. He decided to initiate a more sweeping search effort.

On the twenty-fifth he set Lieutenant Homer and eleven men ashore on the western side of Lukunor Passage with a boat to cross water gaps between the islets. Rather than return to the ship each day, as Strong and Lewis had done, they would move methodically westward along the string of islets, driving the mutineers before them, and ultimately flushing them from cover.

It would be a demanding march, but the men of the *Dolphin* were rested, and they'd

been toughened by their weeks of sailing in equatorial waters. Still, it turned out to be a greater undertaking than Percival had anticipated. When, that evening, they sent a boat out after Lieutenant Homer to refresh his supplies, they were surprised to find that this arm of the atoll "was more than 40 miles in length." The supply boat returned without having found any sign of the second lieutenant.

The full dimensions of the atoll were yet to be discovered, but the strategic possibilities inherent in its topography had become clear. On the day Lieutenant Homer began his search mission, Captain Percival decided to sail the ship through Lukunor Passage into the lagoon. He would be better able to support his shore party from inside the lagoon than from the ocean side. Also, when anchored on the ocean side, there was always the danger of a shifting wind blowing the vessel ashore, whereas the lagoon offered calm waters and easy anchorage.

So Mad Jack lightened the *Dolphin* as much as he could, waited for high water, and attempted to sail her through the cut. However, he'd estimated the depth incorrectly. Strong reports that there wasn't quite enough water in the passage, "and

she grounded without receiving the least injury. We were compelled to heave all aback; get an anchor out astern and haul off." Though neither Strong nor Paulding makes much of it, this was a critical moment for the mission. If they had hung up there, the crew of the *Dolphin* would have been no better off than the men they'd come to remove from the atoll.[5]

On the twenty-sixth she continued sailing west, following the direction of Homer's party. Somewhat concerned as to the whereabouts of these men, Percival dispatched two boats. His own boat sailed along the lagoon side of the islands, and the other, under command of Midshipman Strong, sailed along the ocean side.

Strong found Lieutenant Homer first, and resupplied his party. Then, Captain Percival met up with them and received their intelligence report. The search party had discovered several items left behind by the mutineers, among other things a mitten with the initials of Rowland Coffin on it. However, the distances were great, and the going was rough, so Percival refined his search operation. Homer and Strong were to continue the search as a team, with Strong's boat supporting the shore party.

Unfortunately, after Percival returned to the ship, the weather turned bad again. Strong lost sight of Homer and his men, and was forced to seek shelter in a lee spot on one of the islands. "Having landed and hauled up the boat," Strong wrote, "I ordered two of the men to furnish themselves each with a cutlass (our firearms being wet) and follow me back on the island, to see if there were any natives."

While Strong struggled to reestablish contact with Lieutenant Homer's party, Mad Jack and the *Dolphin* leapfrogged ahead to the most western point of the atoll. Here, on the afternoon of the twenty-seventh, the surgeon died of liver disease, from which he'd been suffering the entire trip.

The crew brought the doctor's remains ashore for burial, much to the natives' fascination. The funeral was conducted with appropriate solemnity until the ceremonial volley of muskets was fired. At this, wrote Lieutenant Paulding, the natives "burst into loud shouts and laughter, for which we drove them back, with threats of punishment." Just a few days before, Paulding had come ashore and unthinkingly shot a white crane that he'd seen outside one of the villages. On learning that the cranes were domesticated, and held in great esteem by the villagers, he expressed regret for his actions. Such were the difficulties of these early encounters, each side behaving stupidly, thinking the other stupid.

By the next morning, Homer's men had again caught up with the ship. After receiving refreshment, the land party crossed a long reef and approached the broadest island yet, apparently the site of a large settlement. Homer and his men, with support from the *Dolphin*, had moved from Lukunor down the long islands of Enajet and Arbar, and were now approaching Mili-Mili, the largest island on the atoll.

Somewhere near the eastern end of Mili-Mili, Lieutenant Homer's landing party came upon staves, pieces of rigging, and scraps of canvas. "Upon digging," wrote Strong, they "found the skeleton of a man, which, they supposed to be that of Comstock, the Chief Mutineer." The men of the *Dolphin* had finally come upon the site of the *Globe*'s landing.

As the trail got warmer, pressure on the searchers increased. Strong says the search party proceeded a few miles farther and encountered a large, agitated group of natives armed with spears, clubs, and stones. Homer didn't like the looks of them. He retreated to the *Globe*'s landing spot and sent for help. Mad Jack responded by dispatching Lieutenant Paulding in a launch with twelve armed men.

By that evening, Paulding and his crew had found their shipmates, none the worse for wear though "looking for our appearance with the greatest anxiety." Lieutenant Homer and his men had spent the last four days under the equatorial sun, feeling their way blindly into a potentially deadly situation. The sight of reinforcements must have been a welcome one.

After resupplying Homer, Lieutenant Paulding sailed ahead to search farther down the atoll. Since it was night, he says, "I suffered the men to lay down and refresh themselves with sleep, while midshipman S. and myself steered the boat."

Alas, this does not mean that Midshipman S(trong) accompanied Lieutenant Paulding on his expedition down the lagoon. Strong's Washington journal makes no mention of his being in Paulding's party, and the Vevay journal notes that when he first came to the *Dolphin*, one of the officers already aboard was a Midshipman Edward Schermerhorn.

The *Dolphin*'s deck log confirms that this other midshipman was the one aboard Paulding's boat.

It would have been nice to place Strong, rather than Schermerhorn, in Lieutenant Paulding's boat, because the next day Paulding discovered the survivors of the mutiny on the whaleship *Globe*.

chapter thirty-three

Survivor

Mili Atoll, November 30, 1825

In response to Homer's call for help, Mad Jack had concocted a plan. Paulding was to sail to the northern arm of the atoll and begin working his way back west. Homer, in turn, would head west and north toward Paulding's party. Both would be supplied and reinforced from the *Dolphin*, waiting at the western end of the atoll, between Homer and Paulding. The two advancing parties would drive the mutineers from cover somewhere near the ship.

Paulding and Schermerhorn left Homer's party that night. While their men napped, they started sailing down the lagoon on a favorable breeze, only to discover that an unfavorable current was actually carrying them back toward shore. They woke the crew, broke out the oars, and rowed until daylight, when they again made sail to the northeast. As they were approaching the northern shore of the atoll, two canoes full of natives armed with spears and stones came toward them.

It was at this point that the true nature of their situation dawned on Paulding. The native canoes could cut across the inside of the lagoon, while the *Dolphin* had to sail its circumference. Compounding the advantage afforded by geometry, the islanders' craft could out-sail Paulding's boat "at least three miles to my one." He now realized that, under such a handicap, he would never apprehend the *Globe* mutineers. They'd simply cut across the lagoon every time he approached. His only option would be to seize the native fleet, piece by piece. The natives would oppose him. It would be war.

With this sobering insight, Paulding approached one of the islands on the northern side of the lagoon.[1] There, he was surprised to find the beach crowded with hundreds of natives, and anxiously noted they'd sent their women and children to the huts. They, too, were thinking warlike thoughts.

He dropped an anchor and was working to keep the boat pointed into the surf, when to his amazement one of the natives began shouting to him in English. The man was standing on a beach about forty yards away from the boat, with the main gang of natives another forty yards behind him. The man was shouting a warning: "The Indians are going to kill you. Don't come on shore unless you are prepared to fight."

Lieutenant Paulding realized they'd discovered one of the *Globe*'s men, though his wild appearance "seemed like an illusion of fancy." His long hair was tied in a knot on top of his head. He wore a loincloth and was so deeply tanned that he almost looked like one of the natives.

The stranger repeated his warning and told Paulding he'd convinced the natives to set a trap for them. He was to lure the boat's crew ashore where they would then be overpowered. The man's story seemed reasonable enough, but Paulding was full of doubts. Could this man be a mutineer? "Why does he not now fly to us for protection, if he is innocent? — forgetting that our contemptible numbers precluded all idea of safety to him."

The young man identified himself as

William Lay and, beneath the tan and the native garb, was a close enough match with the description Paulding had of him. "I then directed him to run to us, but he declined, saying, that the natives would kill him with stones before he could get there. During all this time they thought he was arranging their plan for us to come on shore."

Lieutenant Paulding saw no alternative other than direct action, and thus proved himself a true disciple of Mad Jack Percival. In what a fellow crewman, Charles Davis, called "the boldest act he ever witnessed,"[2] Paulding brought his boat ashore and, in front of the astounded natives, strode up the beach, grabbed Lay with one hand and snapped a pistol at him with the other. "I was not insensible to the sentiment my harsh reception was calculated to inspire; but circumstanced as I was, I could not risk every thing in preference to inflicting a momentary pang, keenly as it might be felt." Paulding demanded the obvious, "Who are you?" to which Lay replied, "I am your man," and burst into tears.

After nearly two years awaiting rescue, Lay became delirious at the prospect of his deliverance, overwhelmed by a very understandable confusion of emotions, babbling half in English and half in the native tongue. The islanders, meanwhile, had moved down the beach, and were now threatening to overwhelm the American sailors in quite another way. Paulding ordered Lay to tell the natives that if they advanced they would all be shot.

This warning had the desired effect. All the natives retreated except for one old man who advanced, despite Paulding's threats. Lay explained that this was his benefactor, "the person to whom he was indebted for life." The old man hugged Lay and wept in a touching manner, but Paulding was getting nervous. The natives had resumed their menacing behavior and seemed to have realized what Paulding knew all along. Once the first volley of pistol shots was fired, the men would be set upon and murdered before they could reload. He hustled his party into the boat, and cast off.

In the calm waters of the lagoon, William Lay told Paulding that the mutineers Payne, Oliver, and Lilliston, as well as the innocent Kanaka Joe Brown, Columbus Worth, Rowland Jones, and Rowland Coffin "were all dead except Cyrus M. Huzzey . . . who was on an island a few miles to the windward of us, and [which was] now in full sight."

Paulding sailed Lay and his crew right down to Hussey's island,[3] and the pistols came out again. Strong relates, "As soon as they arrived at this place Lt. P. presenting a pistol to the breast of the old man with whom Hussey lived, threatening him through the interpretation of Lay with instant death if he did not immediately produce Hussey."

As it happened, Hussey was in the jungle when Paulding arrived. He heard the people shouting for him, but thinking they wanted him to do chores, he did not respond until he heard the repetition of the native word for white man. "He came forth," wrote Paulding, "not having had the least intimation of the approach of those who proved to be his deliverers." The unsuspecting Hussey, in his loincloth and brown skin, presented a remarkable appearance to his rescuers.

Where Lay was nearly delirious with joy, Hussey was in shock: "A phlegmatic stupor together with a seeming bashfulness," was how Strong described it. As with Lay's protector, Hussey's native family seemed genuinely sorry to see him go, and reacted to his departure more like parents than slave owners.

Back on the ship, the two young men were fed, shaved and shorn, and given clothing by the crew. Captain Percival, says Strong, "feelingly shook them by the hand with tears of sympathy for their trying scenes and hardships." Over the next few days, they recounted for the eager crew details of their lives on the island and of the days leading up to their rescue. Augustus Strong wrote it all down, continuing his role as omniscient narrator, and his words, preserved in the Vevay journal, are the first record of those events.

Gorham Coffin's crusade on behalf of his nephew Rowland had been in vain. The application of justice demanded by the merchants of Nantucket had no object. The investigation by the United States government was moot. The great question had been answered. Of the nine *Globe* crewmen stranded on the island, only William Lay and Cyrus Hussey had survived.

chapter thirty-four

The Massacre

Mili Atoll, February, 1824

Twenty-two months before the rescue of William Lay, on the evening of February 17, 1824, Gilbert Smith and his five shipmates made their escape to freedom. Silas Payne, John Oliver, Thomas Lilliston, and perhaps Rowland Coffin — the men who weren't in on Smith's plan — ran down to the water's edge, alerted by the lookout's cry that the *Globe* was gone. At first they tried to convince themselves that the ship had dragged her anchor or somehow drifted away. But the next morning's empty ocean dashed these hopes.

In the narrative that he and Cyrus Hussey published in 1828, Lay writes, "Payne in a paroxysm of rage, vented the most dreadful imprecations; swearing that could he get them once more in his power, he would put them to instant death." One can imagine that poor, dumb murderer hopping up and down on the beach in front of the ship's deserted anchorage. Nothing would bring the *Globe* back to him; his own poor judgment had allowed her to escape. With her, he must have known, went his best chances.

Such was not the case for William Lay and his friend Cyrus Hussey. They understood immediately that if Gilbert Smith survived his voyage, help would be on the way. It was now possible to imagine the end of the rule of Payne and Oliver. However, the two friends didn't feel safe sharing their feelings with any of the others, "fearing they might not agree with us in opinion, and we had too good reason to believe, that there was *one,* who although unstained by blood, yet from his conduct, seemed to sanction the proceedings of the mutineers."

This was an oblique swipe at Rowland Coffin, who they suspected was an informer, but it also paints an unflattering picture of relations among the remaining crew members. Had the violent events of the past month so traumatized Lay and

Hussey that they felt they could not even trust the other teenagers, Rowland Jones and Columbus Worth?

The natives immediately noticed that the ship was gone and crowded around the tent, wanting to know what had happened to her. Payne told them the wind had blown her to sea, and because she had no sails or other supplies aboard, she would be lost. Although this seemed to satisfy them, Payne could not lie as easily to himself. He knew the men had taken her on purpose and that there was at least a fair chance they'd get word back to the civilized world. If he and Oliver stayed where they were, they'd eventually be apprehended and hanged.

So he set the crew to work tearing one of the two remaining whaleboats apart.[1] It was his plan to improve the second boat by building a deck on her and reinforcing her with materials from the first boat. He told the crew that they'd need a seaworthy vessel if the natives turned against them. But it must have been obvious to everyone that nine men and supplies for an indefinite time would be a tight fit in one whaleboat.

The activities of the white men fascinated the islanders. They crowded around the improvised boatyard and continued to frequent the crew members' camp, trading breadfruit, coconuts, and other food items for pieces of iron hoop, nails, and stray bits of white man's culture. Lay says, "Their deportment toward us continued to be of the most friendly nature."

He and some of the other crew members were given leave to explore the islets that composed the atoll. Like Lieutenant Homer's party, they were able to walk a good part of the way around the ring of islands. They headed east over a chain of sandbars, which they called a "causeway," to the island of Arbar. There they came upon what was probably the same village Samuel Comstock had encountered when he left the *Globe*.

Some of the other crewmen then went to the village and shot off muskets, much to the distaste of the inhabitants. "The reader will no doubt agree with us, when we pronounce this to have been a bad policy, for they certainly disliked to have visitors possessed of such formidable weapons."

A day or two after his first visit, Lay returned to the village, and this time he met an elderly man. The two made friends, and the old man convinced Lay to spend the night with his family. "The natives continued in and around the tent until a late hour, gratifying their curiosity by a sight of me. I was provided with some mats to sleep upon, but the rats . . . prevented my enjoying much sleep."

This friendliness was in stark contrast to the conduct of the mutineers.

On Lay's return to camp, and perhaps inspired by his adventures, Payne and Oliver set out in one of the boats to explore the ring of islands. They were gone overnight, and when they returned they brought with them two young native women "whom Payne and Oliver took as their wives." The women passed a day and a night with the mutineers and appeared "much diverted" by the novelty of it all. But on the next day, Payne discovered that his "wife" was gone. He, Oliver, and Lilliston set out with muskets, terrified the inhabitants of the nearby village in a predawn raid, and found the escaped woman. "At this moment one of them fired a blank cartridge," wrote Lay, "which frightened the natives in such a manner that they left the hut and fled. Payne then pursued after, firing over their heads till he caught the one he wanted."

They dragged her back to their camp,

flogged her, and put her in irons. This brutal act sealed their fates.

The next day it was discovered that a tool chest had been pried open, and that a hatchet, a chisel, and some other items stolen. Payne vented his wrath on the natives, who now were crowding the tent more than ever. Eventually, one of them brought him half of the stolen chisel. As a reward, Payne clapped him in irons and kept him chained near the tent all night.

In the morning he selected Thomas Lilliston, Rowland Coffin, Cyrus Hussey, and Rowland Jones and armed them with muskets charged with bird shot. Placing the prisoner in their custody, he sent them into the village to demand that the rest of the stolen goods be restored. The prisoner was supposed to point out the man who had stolen the tools from the chest.

The natives returned the hatchet, but then surrounded the men and began bombarding them with stones. Presumably, Lilliston and company discharged their muskets, but being loaded with bird shot, the guns had no serious effect. The natives continued their advance, ever more menacing, and the four white men began to retreat. Their uncertainty fueled the courage of the islanders, who pressed their advantage. The crewmen broke ranks and fled. Young Rowland Jones, at the rear, was overtaken and stoned to death.

The bruised survivors arrived at the tent in a state of panic. One of their number was dead. The natives meant business.

Because of the natives' ability to sail swiftly from one part of the atoll to another, a few hours saw many more than the "twenty or thirty families" in the nearest village swarming around the tent. "No time was lost in arming ourselves, while the natives collected from all quarters, and, at a short distance from the tent, seemed to hold a kind of council. After deliberating some time, they began to tear to pieces one of the boats."

Payne, realizing that the boats were his only means of escape, took the gamble of attempting to negotiate with the natives. As he went out from the tent, one of their chiefs came forward to meet him, and the two disappeared into the midst of the crowd of islanders. It was a risky venture, demonstrating great nerve and physical courage. Despite his brutality and ignorance, the mutineer apparently had a sufficiency of these qualities. However, his greatest ally was desperation. If he couldn't treat with these natives, all was lost.

After nearly an hour, Payne returned to the tent. He announced that he had come to terms with his antagonists. They were to be given everything belonging to the *Globe*. Payne and his men would agree "to live with, and be governed by them, and to adopt their mode of living."

Lay says that on hearing this news, the innocent men felt themselves to be in a truly desperate situation, through no fault of their own, trapped between the club-wielding savages and "our more than savage commanders."

In fact, they were all in a truly desperate situation. Silas Payne had given ground at a strategically inappropriate moment. The natives gained encouragement from this, as they had from the retreat of the four white men a few hours previously. Although Payne needed to buy time right then, in hopes that he might be able to repair the whaleboat and make his escape, his total capitulation was fatal.

No sooner had Payne announced the terms of his treaty than the natives moved onto the campgrounds and then into the tent itself. The eight surviving members of the *Globe*'s crew watched, helpless and terrified, as the natives began appropriating whatever they pleased. When they began pulling the tent down, Lay says, "An old man and his wife took hold of me, and after conducting me a few rods from the tent sat down, keeping fast hold of my hands." It was the villager in whose hut Lay had spent the night.

After his rescue, safe aboard the *Dolphin*, his words all a-tumble in the emotion of the recollection, William Lay recounted the ensuing slaughter for Augustus Strong, who recorded every word.

Silas Payne, Lay said, was taken some distance away and killed. Then the natives set upon the rest of the crew with spears, clubs, and stones, "and bruised their heads as we are accustomed to bruise the heads of serpents."

One young lad of about fifteen years of age was attempting to make his escape when an old woman plunged a spear . . . into his back. He fell lifeless, but after a moment recovered strength sufficient to raise himself upon his knees. In this posture, with tears running from his eyes, he in the most supplicating voice implored her to spare his life, but in vain! She raised her spear. Distorting her countenance in exultation [she] gave the decisive stroke. Numbers immediately ran up pelting his head with stones until his skull was broken his eyes gushed out and his countenance [was] completely disfigured. This unfortunate young man was an interesting, kind, promising youth, by name Columbus Worth.

At this moment, the two old men and their wives interceded for Lay and Hussey and prevented them from being killed with the rest. "They in this manner escaped without being able to assign a reason for this suddenly formed friendship — Lay however stated that the old woman who assisted in his rescue must have become attached to him for defending her when his companions several times attempted to tease and frighten her."

This is as much as the world will ever know for certain about why William Lay and his friend were saved. Lay had defended an old woman who was being taunted by his shipmates. It is not a perfect explanation, because it includes only the white man's speculation, and not the native's true reasons. But it will have to serve. It did ensure that there were at least two witnesses to what happened.

Resistance was as futile as flight. At some point, Lay and Hussey were separated. Lay says Lilliston and Joe Brown fell within six feet of him, and as soon as they went down, the natives caved in their heads with stones.

When the slaughter was over, all the *Globe*'s company was dead — speared, stoned, and mangled. At least, that was what William Lay thought as he sat glumly in the old man's hut that afternoon. Then the natives crowded around him, all talking at once, and Lay assumed that the moment of his death had arrived. Instead, the crowd parted, and "I saw a number of the natives approaching the hut, and in their midst, Cyrus M. Hussey conducted with great apparent kindness."

It had been just one month since the mutiny.

chapter thirty-five

A Map of the End

Marshall Islands, 2001

 In a folder at the Nantucket Historical Association is a map, probably drawn by George Comstock, of Samuel B. Comstock's final day on Earth.[1]

The map shows two pieces of island connected by a thinner strand of beach, so that the whole looks like a misshapen pair of eyeglasses. On one side of the islands is the lagoon, on the other, open ocean. Each island has an area of woods at its center, and the slender bridge of land connecting the two islands is marked, "Here is a passage over that is covered at high water." Symbols are drawn at various points on the map, and the handwritten legend beneath the islands says they denote:

- from where he started after leaving the shore
- where stopped
- supposed him laying in ambush with some Indians to cut off the crew on Paines firing a gun he came out of the bushes & went to the right
- here we see him no more till the next day when he was discovered coming along with a very quick step
- here he stopped & thought to himself shall I pass the Rubicon or not but passed on over this passage making a cross in the sand with his sword
- here he drew his sword & was soon after shot
- hog pen built by Paine for the hogs but I guess he got mistaken

The caption at the bottom reads, "Sketch of the Ship Globe's Isle visited by the Mutineers."

It is a compelling document, and it would have provided the perfect excuse for a visit to Mili Atoll. But no excuses were necessary. After a year's immersion in the story of the mutiny on the whaleship *Globe*, I had already decided to visit "the Ship Globe's Isle." I wanted to walk where Augustus Strong had walked. I wanted to find the site of Samuel Comstock's death. I had a yen to *inhabit* the story.

Of course, I wound up somewhere else.

The twin-prop Dornier droned over miles of empty Pacific, then banked and descended toward Mili. Even at 800 feet the atoll filled our view — a misty rim of green bounded by white, with cornrows of waves, motionless from this height, around its perimeter, the surreal Kool-Aid turquoise of the lagoon in its center, the whole as vast in real life as it had seemed insignificant on paper. Then we were at water level, coming up fast on Mili-Mili, the largest island of the atoll. A strip of close-cropped grass opened in the jungle, and the plane plopped down on the lagoon side of it, bumped to a stop at the ocean side, and turned and taxied back. As we pulled up to the "terminal," a shack under a palm tree by the water's edge, two dozen dark people in brightly colored clothing gathered to see what the plane would disgorge. The door opened and we stepped into a furnace — acetylene sun overhead, jungle all around, the luminous, backlit green of the lagoon in our faces. (What *was* it about that color?) The people moved closer, making startlingly non-English noises. The airplane headed back down the runway and took off. Already soaked in sweat, my uncertainty edging toward fear, I picked up my bags, imagining I now had some inkling of how Augustus Strong felt when he began that first search with Sailing Master Lewis.

I was traveling with my son Brooks, a professional photographer. We flew from Massachusetts to Texas to Hawaii to Majuro — a long day in an airplane to accomplish a journey that would've taken the *Globe* months to complete. Majuro was a gritty, densely populated urban atoll, a sort of ghetto in paradise, about ninety miles northwest of Mili. From this last stop we planned our campaign. We bought our flour and rice and tinned meat, and reserved our seats on the Dornier, and chartered the boat that would ferry us around the lagoon. An intelligent and good-humored young lady at the Marshall Islands Visitors Authority helped us with logistics, securing housing for us on Mili as well as the services of a translator and a cook.

She also set up the first meeting with our primary informants, Chuji and Beverly Chutaro, who lived in a green cinder-block house on Majuro's main road, about half a mile from our hotel. They had a little shaded porch behind their back door, and it was there that we sat and spoke of the situation on Mili Atoll in the 1820s, a topic of intense interest to them both. Mr. Chutaro was a native of Mili.

He was a wiry man in his sixties, brown, but not as dark as most Marshallese, with gray hair that stood up like a brush. A stroke had weakened his left side, but there was a quality to him — a fire almost — that made him seem anything but weak. He was deep. In his wisdom he preferred not to present himself as such, and so he clothed his gravitas in an old man's slaphappy humor.

It turned out that Payne and Oliver had indeed sealed their doom by brutalizing that native woman. Women were the primary owners of property, and therefore had high status. Sometimes they were even queens. Inevitably, according to Mr. Chutaro, they were powerful. His eyes crinkled in merriment. "Years ago, when all the Chief women were alive, they'd have *eaten* you. Oh Boy! But now they're all dead. Good thing, huh?"

Of course, he continued, in those days it was customary for people to kill any outsider who came to their island. So even without Payne's bad behavior, the long-term future of the mutineers was not bright.

As to why Lay and Hussey had survived, Mrs. Chutaro pointed out that adoption was a common practice on Mili; a family could "adopt" the son of a brother or a cousin, but it was a different sort of relationship than just caregiving. Life on the islands was hard, and the more help a family had, the better. Among the Marshallese, all sons, adopted or not, worked at fishing and collecting pandanus and coconut. Both Lay and Hussey's "parents" were elderly. Having a son to help them with these labors would be a tremendous benefit.

Mrs. Chutaro was fair and blond. She'd met her husband when they were at college in the States, but they'd moved back to the Marshall Islands to raise their family. She was an anthropologist specializing in Pacific studies, and it was she who pointed out that whiteness must have been another reason for the survival of the two boys. White skin was a novelty, but it was also considered a mark of beauty. Before marrying, Marshallese women were kept in darkened huts so that their skin would become lighter, and they would be more attractive on their wedding day. Because of their whiteness, Lay and Hussey would have had an ornamental or "pet" value.

Mr. Chutaro chimed in with what he called "the imaginings of a dirty old man" — speculating that, because of their whiteness, Lay and Hussey would have been magnets to all the marriageable girls on Mili, thus enhancing the social status of their parents. He also suggested that the people who adopted these boys must have had some standing in the community to begin with.

In any event, the novelty of their whiteness, the benefit of having a "son" to help with the work, the fact that the "parents" had some standing in the community and could argue for Lay and Hussey's survival — all added to the main reason for their salvation. Lay had "frequently" (the word he used in his narrative) protected the old woman from the taunts of his shipmates. And Hussey was his particular friend.

Mr. Chutaro qualified this opinion with a disclaimer that he repeated several times during our conversations. "These are just my thoughts. My impressions from knowing Pacific and Marshallese culture. *There is no tradition of the event, and there is no*

physical evidence. No s
vived."

I gave him my paperba
and Hussey's narrative, an
one sitting, so compelling w
to him. From his earliest
been the one who asked all
stories and traditions of his p ... , ai
eventually, he became the living repos.
of these stories. He could trace the desce
of people we would meet on Mili back to
the chiefs who had adopted Lay and
Hussey.[2] However, because no trace of the
event had survived in his culture, it was
not until he was in the United States that
he first heard the story of the *Globe.*

That had been many years ago, back in
the sixties. Now he was delighted to renew
his acquaintance with the tale. He knew
from personal experience all the places Lay
and Hussey referred to in their account,
and he drew his own map for me, changing
their 19th-century orthography into the
modern place names, where Dillybun, for
example, became Chirubon, and Tabara-
wort translated to Jobenor.

He clarified the frequent and confusing
shifts of location in Lay and Hussey's narrative by explaining that Marshallese were
great sailors,[3] and that most landowners

409

25 owned property on several
d spent their time commuting
d forth between them.

, armed with Mr. and Mrs. Chutaro's
tural information, George Comstock's
ap, Augustus Strong's account, and the
ther narratives of the mutiny and massacre, Brooks and I marched up and down Mili-Mili in hopes of locating the scene of each event that took place there.

At the northern end of the island, we saw where the *Dolphin* might have anchored in 1825, and sent Paulding into the lagoon. We waded out to the bar between Mili-Mili and Madagai, the island to the north, to satisfy ourselves that the *Dolphin*'s launch and cutter might have passed over it at high tide. We made trips to the eastern end of Mili-Mili, where it narrowed and connected with Arbar. I drew charts based on the narratives, and compared them with George's chart. Then I matched them all against the topography, until I arrived at a best guess of the *Globe*'s anchoring place and the site of the massacre, between Mili-Mili and Arbar.

The thirty-five-foot sport fishing boat that we'd chartered on Majuro arrived after three days, to run us the length and breadth of Mili Atoll's huge central lagoon.

We attempted Lukunor Passage, where the *Dolphin* tried to get in, and just as they had grounded in 1825, we were forced to turn around for fear of getting stuck. We cruised up to the beach on Tokowa where Lay was discovered by Paulding, and then over to the jungle in Jobenor where Hussey hid, thinking they were calling him to work, until he heard the Marshallese word for "white man."

But the atoll was only sand. Its features had shifted many times in the memory of men like Chuji Chutaro, and who could say how many more in 175 years? The texts were imprecise. Locations were relative. A "westernmost" point might seem so until the point was rounded, revealing another ten miles of west-trending island. None of our researches produced anything more than educated guesses. Our mapping and figuring was just an intellectual game. What we really learned was the deeper meaning behind Mr. Chutaro's assertion that no physical remains of the mutineers existed, and no trace of their story survived in the folklore of Mili Atoll.

World War II had obliterated it all.

Gradually, in the course of our week on the atoll, we realized the war was the dominant fact of the place. We could not walk more than a few minutes in the jungle without encountering the smashed hulk of a Japanese "Zero" fighter or "Betty" bomber, or concrete gun emplacements, or the massive guns themselves, rusting like the remains of some nightmare civilization in the remorseless tangle of green. As soon as we ventured off our island's main road (which, we learned, had been back-filled and landscaped), we encountered lunar terrain. Every few feet was a twenty- or fifty-foot bomb crater. There were carcasses of sixty-year-old shipwrecks on the beaches, and traces of bombed-out wharves and jetties. Casings of dud bombs poked through the beach sand like shells of alien mollusks. The long-dead war machines intruded with an ugly insistence, locked in battle with the jungle for dominance of the landscape. The only consolation was that the jungle would win.

Eventually it dawned on me that the great guns of the island could only have functioned if all the vegetation had been cut down. The Japanese would have cleared the northern end of Mili-Mili, and the clearing would have extended to the runways in the wide, central part. The runways, in turn, would have covered the rest of the island, clear to its narrow eastern

412

end. In World War II, it had all been shaved bare.

Mili-Mili had been a depot and administrative center for the Japanese.[4] More than 5,000 soldiers and sailors were stationed there. After 1943, American soldiers fought their way up through the Gilberts, then jumped over Mili and secured Majuro and Kwajalein in the north, cutting off supply lines to the Japanese garrison on Mili-Mili. To soften resistance on the island, the Americans bombed heavily. According to a government report, "This island is saturated with bomb craters and the number is estimated to be anywhere from four to eight thousand. . . . One large crater had about a dozen fishes that resembled mullets. . . . They were reported to be edible. . . . In a way the craters can be referred to as a Godsend."[5]

Two thousand five hundred of the Japanese stationed on Mili died of starvation, suicide, and bombs. Things were worse for the natives.

Mr. Chutaro told me that his father had been Japanese. His mother was Marshallese. When the war came, she'd traveled to another island to try to find a better situation for her family, but the Japanese detained her and ultimately killed her. "And when there was food, guess who got it? Not us, I'll tell you. Those were some rough times, all right."

Payne's savagery was nothing compared to what the Japanese and the Americans inflicted on one another. I got a nasty glimpse of the unintended consequences of America's eager commercial agenda in Mad Jack's day; of the "civilization" the missionaries wished to impart. But this was tempered by something else Mr. Chutaro had told me.

On the eve of our departure from the urban island of Majuro to Mili Atoll, I was still jet lagged. Our preparations, a process of learning over and over how ignorant we were, had been stressful. I had a touch of food poisoning to boot, and was suffering its flu-like symptoms. When I went for my final interview with Mr. Chutaro, I must have seemed too ardent, feverish even, in my declaration that, after all he had done for me, I *needed* now to do something for him, and what could I bring him, or buy him, or cause to be done for him?

He smiled. "We're all the same in God's sight. You're the great writer, and I'm the old man, but I'm no better than you. You're no worse than me. So He will take

414

care of these things, and you don't need to feel like you have to thank me."

I don't know why I was so unprepared for that utterance, but he might as well have turned into an orchid. The smile broadened as he sensed my confusion. Then, probably to reassure me, he told me a story.

He had arrived in the United States in the early 1960s, just as the Civil Rights movement was taking hold. He was being sent, by missionaries, I guessed, to Heidelberg College in Ohio. He'd never been off his little island, and suddenly he was in Honolulu and then Los Angeles and then Houston.

"Oh, Boy!" He laughed. "There was still segregation then. But I didn't know. I didn't know *anything*. I hadn't eaten in a day. There was no time to eat in Los Angeles. So I went to a lunch counter at the airport in Texas." Predictably, he was ignored. When he finally mustered the nerve to show his money to the man behind the counter and ask for food, the man told him flatly he couldn't serve niggers. Young Chuji walked away, disconsolate, and bumped square into a tall man standing at the edge of the dining area.

"Just like John Wayne, I swear. Big tall

guy, cowboy boots. He even had a white hat. No kidding! And he asked me, 'What's the matter, son?' And I told him about the guy at the counter. And he just walked over there in those cowboy boots — stepped right over the counter! And he said something to the guy, and next thing I know I'm sitting in front of a big steak. Boy, was I hungry!"

The plane flew to Columbus, Ohio, and landed in the middle of the night. No one was waiting for him at the airport, and he had no idea what to do. When the sun came up, he left the terminal and found himself in a city surrounded by cornfields. No one could help him get to his college. Desperate, he went to a phone booth, picked up the phone book, and, because he didn't know where else to look, dialed a government number. Next thing he knew, by some absurd miracle, he was talking to Governor James Rhodes.[6] The governor patiently listened to Chuji's story, and then told him to wait right where he was, not to move.

"So I waited. And all of the sudden I hear sirens. And up pulls a police car, and I think, 'Oh no! They're coming to get me. They're going to throw me in jail.' "

Instead, they took him to a hotel, where

Governor Rhodes had booked him a room and left him an envelope containing fifty dollars. "Can you imagine being on an elevator after living on Mili? There was a shower in the hotel room, and I didn't know how it worked. I walked in there all dressed up and turned a knob . . . Soaking wet!"

But he was too wrought-up, too close to his destination to stay in the hotel. He went out and walked to the cornfield. He'd never seen anything like that, and it amazed him as much as the shower or the elevator. Then he found cabs lined up outside the hotel and asked the way to his college. But the college was too far away. No one would drive him there.

Just as he was about to give up, a cabbie approached him. The man had been to the Marshall Islands as a GI during the war. He drove Chuji right to his dorm. Wouldn't even take any money. And that was how he came to America.

Mr. Chutaro leaned back in his chair and beamed at me. I was much calmer now. "So," he said, "people take care of each other. Those two boys, Lay and Hussey, and the great mystery of why they survived? I think it was because God meant them to."

I liked that leap and its implied connection to the two of us. I liked thinking that the pious Lay and Hussey would have agreed wholeheartedly. Seeing me smile, Mr. Chutaro said, "I truly believe that."

"I do, too," I told him.

chapter thirty-six

The Diplomats

Mili Atoll, 1825

According to what Lay and Hussey told the crew of the *Dolphin*, their schooner had been sighted, and the whole atoll informed of its presence, well before it landed. (The fleet of outrigger canoes zipping back and forth across the lagoon took on a different meaning.) The first timid visitors to the ship had been spies, not traders. They carefully noted the armament and the number of men on the vessel, and reported these details to their leaders.

Once Lieutenant Homer's party had been sent out, "every motion of the searching party was observed, and information of their proceedings arrived hourly to the Chiefs." The natives questioned Lay and Hussey about the vessel, but both professed ignorance as to its purpose. However, those native spies made it clear that the *Dolphin* was a ship-of-war, and that things would go badly for the islanders once the navy men learned what had happened to the crew. They decided to behave in a friendly manner toward the white men, then to attack them when their guard was down. "While part of the crew were on shore," wrote Strong, "several of the natives were to swim off to the vessel, take the men when off their guard, and throw them overboard, those that remained in the water were to drown them . . . those on shore were to fall on the watering party and massacre them."

The chiefs had firsthand knowledge of the brutality of Payne and Oliver, and no reason to expect that the men of the *Dolphin* would behave any differently once they learned that their white brothers had been murdered. Strong adds, "This plan was not formed from barbarous inclination merely to rob and murder but they considered it as necessary for their own preservation to exterminate the white men, who they expected would totally destroy them."

As it turned out, on November 24, when the *Dolphin* was anchored at Lukunor Passage, they were directly opposite the village where Lay was being kept. The many canoes pulled up on the beach belonged to local chiefs who'd gathered for a council of war. In fact, when Mad Jack landed, the natives secreted Lay in a hut not twenty-five feet from that beach, with a guard stationed over him, ready to crush his skull if he should make a peep. Percival passed so close that Lay caught a glimpse of his epaulets, and understood then that the ship was an American man-of-war.

Somehow, he was able to take the long view, and remain silent, later to convince the natives of the plan that led to his rescue. Still, when he saw the captain's gig cast off, he must have had a dark moment.

Despite everything Lay and Hussey told about the lives and customs of the natives of Mili Atoll, the terms of their bondage remain as unclear as the reasons for their salvation. Both boys were made to work, and work hard at times, by their respective families. Yet both families seemed genuinely fond of their charges, and Strong's narrative, as well as Lay and Hussey's, contains moving scenes of the two young men parting from their aged rescuers.

Lay and Hussey related the events of the mutiny for Strong, Paulding, and the rest of the *Dolphin*'s crew. They also provided gory details of the massacre and its aftermath (such as the fact that the natives had used the bones of their comrades "for ornaments and useful purposes"). When they returned to Nantucket, Lay and Hussey recounted their adventures, complete with a long Mulgrave-English vocabulary. Their exotic experiences among the savages were at least as big a draw as Comstock's bloody murders, and the two strands combined to make their *A Narrative of the Mutiny on Board the Whaleship* Globe a steady seller for 175 years. Strong's journal provided a major source for *Journal of a Cruise of the U.S. Schooner* Dolphin, written by his friend and mentor, Lieutenant Hiram Paulding.[1]

With the fate of the *Globe* mutineers settled, and the two survivors rescued, Mad Jack and his men now assumed the role of diplomats.

The natives, of course, assumed that their game was up. However, Captain Percival, employing the two young men as interpreters, assured them that no harm would come to them as long as they remained peaceful. Lay and Hussey's native families and friends were given visiting rights aboard ship, rights they exercised with great trepidation.

Then, early in December, Mad Jack sent out word that he wished to meet with all the high chiefs of the island, either aboard the *Dolphin* or in some coconut grove ashore. "This the natives were afraid to comply with, supposing this was the time appointed for their punishment."[2] The captain replied that they'd better show up, or he *would* punish them — "would make the whole land ring about their ears" — was the way Strong recorded it, and one can imagine William Lay standing on the beach translating this picturesque, imperious, and frightening phrase for the native envoys.

They gave a white handkerchief to an old woman, along with instructions as to its proper use, and next morning, out she came waving the white flag. This was answered by a white flag from the schooner. The high chiefs of the island were ready to receive "the Great High Chief" of the *Dolphin*.

Here the story takes a noble turn. As Strong recorded it, Mad Jack came ashore in full uniform. He told the chiefs that because of the character of Payne and Oliver, and because of the "insults and bad conduct of the white men," he would not punish the islanders for the murders they had committed. "At the same time, he told them, had they killed the two innocent ones that were preserved, he should have taken singular revenge upon them, and that hereafter they never must molest white men, who should chance to fall among them. If they did he would know it and return."

He required them to return the *Globe*'s whaleboat, and some other items, which they did. Great High Chief Percival then presented them with cotton handkerchiefs and axes to prove his friendship and seal their agreement. He also offered the services of the ship's carpenter to caulk and repair the chief's great war canoe.

The relief and gratitude these people felt must have been tremendous. Their sincerity is expressed in a disturbing anecdote related by Strong. While the canoe was being repaired, the captain happened to mention, in an offhand way, that he didn't like the looks of one of the native chiefs. The high chief "immediately inquired of him what the Great Chief said." Without reflecting, Lay told him. A few minutes later, the offending chief got up and went out, and "the next day was found a corpse in the woods."

Finally, "the Capt. also presented the Chief with a male & female pig. They were delighted beyond measure. For some time they carried them about in their arms." Whereupon, after a few more days of sight-seeing and ceremonial visits, Mad Jack and his crew departed. However, they did not sail off into the sunset. One more strange adventure awaited them.

As he'd been ordered, Percival sailed to Hawaii to look after the interests of American merchants there. Ultimately, he convinced the Hawaiian chiefs to acknowledge their own debts and the charges rung up by Kamehameha II. They even agreed, in general terms, to repay them. Thus, as his biographer cheerfully notes, "Percival merits a tiny historical footnote as the creator of the Hawaiian national debt."[3]

Mad Jack also took action to reduce the population of deserters and beachcombers that had produced the likes of Payne, Oliver, and Thomas. The American agent in the Sandwich Islands wrote in support of Percival, citing the salvage of an American ship, Percival's timely action in preventing a whaling vessel from running ashore in a gale, and an unspecified "immense amount of property" saved by the men of the Dolphin.[4] Still, when Percival and his men set sail for South America on May 11, they departed under a large cloud.

By his aggressive and arrogant behavior, Mad Jack had managed to make enemies of several of the merchants and all of the American missionaries at Honolulu. Two traders named Edwards and Sistair got in a series of disputes with Percival, one of which came to blows. Worse, the crew of the Dolphin, encouraged by Mad Jack's attitude and assisted by like-minded whalemen, so strenuously objected to a missionary taboo placed on local prostitutes that they rioted in the streets. Arch-missionary Hiram Bingham narrowly escaped a beating, and for the next decade, outraged missionary factions in America tormented Captain Percival.

For dealing with savages and capturing mutineers on the remote Mulgraves, Commodore Hull couldn't have picked a better man. When it came to sorting out the situation in the Sandwich Islands, Uncle Isaac should have gone himself.

chapter thirty-seven

The Almanac

Nantucket, 1827

Generally speaking, it is unwise for book-sellers to collect the items in which they deal. Aside from my reference books, I own no rare books or artifacts pertaining to maritime history — with one exception. Years ago, I found in a New England auction a small lot of *Farmer's Almanac*s from the 1820s. At the top of the front cover of the 1824 issue the name "Cyrus Hussey" was inscribed. This caught my interest, and, since I'd already been introduced to the story of the *Globe* mutiny, I looked through the old yellowed booklets to see if there was any evidence that their owner had been *the* Cyrus Hussey.

This was not the case. These almanacs,

occasionally inscribed with such rustic notations as, "cow went dry" or "very cold," had not belonged to Cyrus M. Hussey of the whaleship *Globe*. They had belonged to his father, waiting back on Nantucket, hoping that his son was one of the few rumored to have survived. Written in the margin of the page for October 1824 was the comment "News of the Globe mutiny." On the next page, November, "Globe arriv'd at Edgartown" and "This day Globe here. Glbt. Smith took tea with us & spent the evg."

Then on the page for April 1827,

> Recvd letter from my son C M Hussey 29–4–1827 dated at New York on Board of the Ship United States Commodore Hull.

The boy was alive! He'd returned aboard the *Dolphin* to Valparaiso the previous July, and in January 1827, when Isaac Hull's tour of duty was up, he'd come home with Lay, Percival, and Uncle Isaac in the frigate *United States*.

News of the survivors had beaten Cyrus Hussey back to America by a considerable margin. On August 26, 1826, the *Boston Patriot* reported that the *Dolphin* had arrived in Woahoo with the survivors of the mutiny, but it did not name these men. The lack of detail regarding the identity of the survivors must have been maddening to Cyrus senior and others awaiting word. In short order, Gorham Coffin fired off a letter to Secretary Southard requesting more information. Southard replied that there was nothing further to tell at that time, and no news was forthcoming until several months later, when the whaleship *Loper* returned from the Pacific. Her captain had spoken with Lay and Hussey in Hawaii. On October 28, 1826, the *Inquirer* was able to print the grisly summary, albeit with some inaccuracy.

Thus, of the ship's company, which consisted of 21 persons, one third were massacred by the mutineers; one third were destroyed by savages; and of the residue, five have lived to return in the Globe; the other two, Cyrus M. Hussey, of this town, and William Lay of Connecticut, are accounted for. . . .

For six months, from October 1826 until April 1827, this was all Cyrus Hussey's father had to pin his hopes on. The story was settled with finality when the *United States* arrived in New York on April 21, 1827. Young Cyrus must have sent his letter home as soon as the ship docked.

Unfortunately Mad Jack Percival's enemies also beat the *United States* home. His old nemesis, Captain Edwards, filed assault charges in New York, and, instead of being welcomed as a hero when the *United States* docked, Mad Jack was arrested. Eventually, Percival was found guilty and fined $100. The adverse publicity was accompanied by well-placed attacks from the missionaries in national newspapers. Their lobby created sufficient political pressure that the navy was forced to hold a court of inquiry in 1828. By this time, Percival's enemies had assembled a great mass of documentation, and charges against him ranged from allegations of personal immorality to assertions that he had interfered with the Hawaiian government. He was ultimately found not guilty on all counts, but this decision was widely regarded as a political whitewash. Percival's career was tainted (or given extra color) thereafter. He made the rank of captain in 1841, retired in 1855, and died in 1862.

Hiram Paulding rose to the rank of rear

admiral and continued to serve his country through the Civil War, playing a role in the construction of the *Monitor*, and helping to protect lives and property during the infamous New York Draft Riots of 1863.

The trail of Augustus Strong, a lesser figure in the pages of history, is more recondite. It led back to Indianapolis and Jay Small's protégé John Mullins who, from his post at the Blue and White Truck Stop, directed me to Vevay, Indiana. In the assessor's office there I learned that Theodore Langstroth of Cincinnati had purchased an old church that he'd filled with books, paintings, antiques, and other treasures. Upon his demise in 1978, the books, Augustus Strong's journal among them, passed to Jay Small. I also learned that the books, and Langstroth himself, had come from Cincinnati, where Augustus Strong's family had been prominent throughout the 19th century. This took me to the Cincinnati Historical Society where a large trove of Strong family papers was stored. According to those documents, Augustus Strong was serving as first lieutenant aboard the U.S. schooner *Experiment* at Pensacola, Florida, when he died of yellow fever in 1834. Sadly, he'd been just about to make that long-contemplated voyage to the "land of matrimony" when death snatched him. His letters revealed that the duel against Bushrod Turner in the probable "work of fiction" had been real; it had been averted at the last minute. More surprising, Augustus R. Strong had only been seventeen years of age when he wrote the journal of his cruise on the *Dolphin* in search of the *Globe* mutineers.

Lay and Hussey, who'd done well under Uncle Isaac's care, were transferred from the *United States* when she landed at New York in 1827, and placed in the custody of a Nantucket man named Macy.[1] Because they were possible accessories to a capital crime, their status was dependent on the courts. Within the month, a judge in New York released them on their own recognizance, and the two young men went home.

The story of what had happened to them aroused such interest that they were encouraged to publish an account of their adventures. The project was probably initiated by William Lay. His is the main voice in the book, and he was the one who found a printer in New London, Connecticut. It also seems he found a ghostwriter to help shape their tale. By September he was writing to this unnamed individual.[2]

Do get the work revised and come on here with it as soon as possible and don't delay. I want to get through with it this fall and get the books distributed. I shall expect to see you on here with the narrative as soon as the 15th of October if not before. If Cyrus H has gone to sea I shall expect his father to come on just the same and we shall proceed the same and get the work done.

Hussey was indeed at sea. Just three months after his return to Nantucket, he signed on the whaleship *Alexander* as cooper. However, his experience on Mili had fatally weakened him. He became so debilitated by consumption that he was sent home on the ship *Congress*, and he died before the ship reached America. He was twenty-two years of age. Lay moved to Southport, Wisconsin, where, in 1848, he issued a revised edition of his book. The editor of this second edition refuted the theory that the Lay and Hussey narrative was ghostwritten. He stated that Lay had composed his own account in 1824, consisting of "113 closely written pages of foolscap folio," and that this new edition was based on it.[3] In either guise, *A Narrative of the Mutiny on Board the Whale Ship*

Globe quickly diffused into the American imagination.

Over the next century, one aspect of the *Globe* saga dimmed and warped into a caricature of history. By December 20, 1942, our naval forces were engaged in bitter fighting with the Japanese for control of the Gilbert and Marshall archipelagos, and islands like Mili Atoll were just reentering American consciousness. A striking photo appeared in the *Boston Herald* that day, but it was not a picture of airplanes or soldiers. It showed a group of outlandishly got up natives — right from central casting, with grass skirts, garish body paint, and bones in their noses — standing behind a heap of human skulls. The caption read, "Collecting skulls was always one of the hobbies of the natives of the little island of Mili."

A few traces of the *Globe*'s crew can be found on Martha's Vineyard and Nantucket today. Documents in the probate court on the Vineyard attest that Third Mate Nathaniel Fisher, who died at the hand of Samuel Comstock, left an estate consisting of four pairs of trousers, two shirts, one dress shirt, a pair of drawers, and a keg. The total value of these items was $8.50.

Thomas Worth's estate was more substantial, amounting to $2,284.67. The portrait doesn't show up in the inventory of his personal effects, but an accompanying document requests that his widow, Hannah, be allowed to keep "a watch & portrait of her late husband." Papers in the same file show that "Peter Kidder & Brother" repaid Hannah their debt of $1.75.

The Kidder brothers went back to sea. Peter died in a whaling accident in 1827. Stephen wound up back aboard the *Globe* some years later, serving as first mate.[4]

Gilbert Smith, the young man who masterminded the escape, was rewarded by the Mitchell partners by being made master of the *Globe*.[5] However, within a short time, he left Martha's Vineyard for France, where he raised a family and spent the rest of his life. This seemingly whimsical decision makes sense in the context of whaling history. The French, always seeking to enhance their whaling industry in the face of competition from England, had for decades lured Nantucket specialists into its fleet. In the 1790s, a colony of about a dozen Nantucket whaling families was established in Dunkirk under the leadership of the brilliant Quaker businessman

William Rotch. Gilbert Smith's move to France had its roots in this relationship. He never left whaling.

Some sperm oil — the original purpose of all this — had been brought back with the *Globe*, and the Mitchell partners paid out shares to the living and dead alike. Fisher's estate received $107.18. Captain Worth's share was $264.62.

Rowland Jones's mother and stepfather set a gravestone in his honor at the West End Cemetery in Edgartown. The stone has weathered, but the writing incised in it, and the emotions behind it are still clear:

Far far from home from
Friends & Kindred
By savage hands this lovely youth
was slain;
No fathers pity or no mothers tear
Soothed the sadness or eas'd the
hour of pain

Twenty miles southeast of the cemetery, in the probate court on Nantucket, a sheaf of papers document Gorham Coffin's death from consumption in 1849. He remained a solid citizen to the end, continuing to invest in vessels, and playing an important role in support of the Coffin

School. But by then the Nantucket branch of the whaling industry had lost much ground to New Bedford. Gorham's investments were difficult to liquidate, and he died in debt to the Coffin School.

Along with George Comstock's map, several other artifacts are stored in that folder at the Nantucket Historical Association. Chief among these is the narrative George wrote out for Gorham Coffin. A Coffin descendant prepared a typescript of this account, and this is in the folder, too. At the end of the typescript, the copyist added a note tracing the provenance of the manuscript back through the family to Gorham himself, thus lending credence to the theory that George wrote his account for the Mitchell partners to provide information about the mutiny. The same note says that George went to San Francisco, and that in later years he saved the life of a youngster whose sailboat capsized in the harbor there.[6]

The *Globe* herself performed a final whaling voyage and then was sold to the merchant trade in South America. Probably she met her end on those shores, but there is one tantalizing shard of a story in the folder of papers relating to George Comstock.

In the early days of California there were moored along the bank of the river at Sacramento a number of old hulks, which served the purpose of docks, and among them there was seen by a son of one of the original owners the remains of the ship Globe.

The notion that Gorham Coffin's son saw the *Globe* in Sacramento is just barely within the realm of possibility. A stoutly built whaleship could survive thirty years or more, and old hulks were often used as fill by rapidly growing cities. The idea is as romantic as it is implausible, and it should stand thus, as much a testament to the unstoppable trajectory of the tale as to the handiwork of the old shipbuilder Elisha Foster, the workers who assisted him, and his stout New England oak.

Key to Chapter Silhouettes

Chapters 6, 12, 18, 24
Making a passage — under fore and main royals and stu'ns'ls, on course from Cape Horn to Hawaii

Chapters 1, 7, 15, 19, 23
Making a passage — Hawaii to Fanning Island

Chapters 2, 8, 16, 20, 26
Cruising on the Japan Whaling Grounds

Chapters 3, 9, 13, 21
Hove-to — Boats lowered for whales

Chapters 4, 10, 14, 22
Cutting in — Topsails and jib set to make headway

Chapters 5, 11, 17, 25
Under shortened sail — Escape from Mili to Valparaiso, February 17–June 6, 1824

Chapters 27–37
United States naval schooner *Dolphin*

Silhouettes by Erik Ronnberg

Notes

Prologue
1. See "Report of Capt. Shubael Chase," *Nantucket Inquirer and Mirror*, April 18, 1822.

Chapter Two: Quixote Bent
1. In 1801, the North and South Meeting Schools joined into a single institution. See Leach and Gow, *Quaker Nantucket*, pp. 155–56.
2. William Comstock, in his *Life of Samuel Comstock*, says Samuel's mother left a family of "eight small children" on her death in 1818. However, *Nantucket Vital Records* lists only five children born to the couple, and two more born to Nathan's second wife.
3. Leach and Gow, *Quaker Nantucket*, p. 156.
4. *Longworth's Directory* lists Nathan Comstock at Bowery, Oliver, Lom-

bardy, James, Catherine, Front, and Market Streets between 1812 and 1822.

5. Fairburn, *Merchant Sail,* vol. 1, 504–05. Richard J. Cleveland, in his more detailed account, admits only that they were carrying a cargo "consisting principally of European manufactures." According to him, the *Beaver* remained captive until November 1818. Cleveland spent two more years with the *Beaver* on the South American coast and ultimately made $75,000. Cleveland, *Voyages,* pp. 284–342, 402.

6. Starbuck, *History of the American Whale Fishery,* pp. 220–21.

7. Creighton, *Rites and Passages,* pp. 9–12, citing the journals and letters of Robert Weir, Nathaniel Robinson, and Charles Stedman. "These men were the Richard Henry Danas of the ship's company who, by virtue of their 'talent and information' (or, as we might put it, education and wealth), dissociated themselves from other sailors."

8. Leach and Gow, *Quaker Nantucket,* p. 7. The "oil and ignorance" quote is from Comstock, *Life of Samuel*

9. Beddie, *Bibliography of Captain James Cook*, p. 656.

10. Archibald Duncan's *The Mariner's Chronicle*, a very popular anthology of shipwrecks, was first published in six volumes in 1804. American editions appeared in Philadelphia in 1806 and 1810. See Huntress, *A Checklist*, p. 143C.

11. Stackpole, *The Sea-Hunters*, p. 296.

12. Wilkes, *Narrative of the United States Exploring Expedition*, vol. 3, 47.

Chapter Three: The Island and the World

1. Obed Macy, *History of Nantucket*, pp. 41–50, recounts these early developments.

2. Tower, *A History of the American Whale Fishery*, chapter 8, discusses whale oil in commerce.

3. Crevecoeur, *Letters from an American Farmer*, p. 115. There were four wharves by 1801, five by 1811, reflecting Nantucket's growth.

4. Kugler, "The Penetration of the Pacific by American Whalemen in the 19th Century," p. 23 in *National Maritime Museum Maritime Monographs and Reports, no. 2 (1971)*.

5. Obed Macy, *History of Nantucket*, p. 79.
6. Ibid., p. 80.
7. Crevecoeur, *Letters from an American Farmer*, p. 114.
8. See Obed Macy, *History of Nantucket*, pp. 95, 122–25.
9. Starbuck, *History of the American Whale Fishery*, p. 90.
10. Starbuck, in *History of the American Whale Fishery*, notes that the *Industry*, which sailed from Nantucket the same year as the *Globe*'s fateful trip, was "manned wholly by blacks." Starbuck, *History*, pp. 242–43. Particularly in the early years of the 19th century, blacks had a chance of advancing to an officer's berth. See also, Stackpole, *Sea-Hunters*, p. 287, and Nathaniel Philbrick's excellent treatment of the matter in *Away Off Shore*, chapter 16, especially pp. 180–81.
11. Obed Macy, *History of Nantucket*, p. 182.
12. Porter was an accomplished writer and, happily, left us an excellent account of this cruise. Published in 1815, it is entitled *Journal of a Cruise Made to the Pacific Ocean*. Speaking of

this narrative, historian Samuel Eliot Morison calls it "the best bit of sea literature of the period." Morison, *Maritime History*, p. 203. (Although Morison's scholarship has been called into question by modern historians, his taste and style are beyond reproach. His *Maritime History* is still fun to read.)

13. Stackpole, *Whales and Destiny*, p. 350. Stackpole, too, has been challenged with regard to accuracy. But he was the first man to cover the ground, and deserves respect for the amount of material he uncovered.

14. Porter, *Journal*, vol. 2, 78–9. Neither the name nor the claim stuck.

15. This is conjecture, based on the fact that Paul Mitchell signed her 1815 certificate of enrollment. Starbuck does not list an owner until the *Globe*'s third trip, when the owner is given as "P & C Mitchell" (Starbuck, *History of the American Whale Fishery*, p. 232), suggesting that ownership was shared with his brother Christopher. By the time of her fourth voyage under Thomas Worth, the *Globe*'s ownership had gone over to Christopher Mitchell & Co. (Briggs, *History*

of Shipbuilding, p. 231; Starbuck, *History of Nantucket*, p. 498). In fact, the issue of ownership, as opposed to management, is murky. While a company like P & C Mitchell would have overseen the construction and outfitting of a new vessel, her actual capitalization — and hence the risk of ownership — would have been divided among many parties, from the partners themselves to other professionals, such as the captain of the vessel, to the proverbial widows and orphans of Nantucket. Furthermore, the boundaries of these shipowning companies were often indistinct. Paul Mitchell operated a firm that owned the *Foster* and several other vessels. His brother Christopher was the titular head of a company composed of sons Seth and Charles, and son-in-law Gorham Coffin. They probably all had a financial interest in the *Globe*.

16. Of this excitement, Macy says, "Business was commenced with alacrity. . . . Believing that the first oil imported would command a high price, many were stimulated to embark in business, beyond the extent of

their funds. Thus a system . . . of long credits was introduced, which, though it promised some advantage . . . was nevertheless pregnant with evils." Obed Macy, *History of Nantucket,* p. 205.

Chapter Four: The Birth of the *Globe*

1. Obed Macy, *History of Nantucket,* p. 206. Macy notes record cold in February 1815.
2. According to the Yankee chronicler responsible for recording such things, Elisha Foster was a "heavy, stout man" who had four sons in the business with him. This same historian says that Foster's friends advised him against becoming a shipwright, because timber was so far exhausted that his business would soon fail. Briggs, *History of Shipbuilding,* pp. 228–41 summarizes the careers of the various Foster shipbuilders.
3. Nantucket probate records indicate Coffin owned shares in Boston and Cambridge insurance companies. See book 18, Nantucket Probate Court Records.
4. See Baker's *Sloops & Shallops,* pp. 142–52, for details and drawings of a

North River sloop of that era, as well as a discussion of the river and the shipbuilding that took place there.

5. Some of them were of enduring fame, such as the *Beaver*, from whose decks the Boston Tea Party took place. Shubael Chase's *Foster* was built along the river as well, and named after Elisha.

6. Other than a few details mentioned in Briggs's *History of Shipbuilding*, pp. 231–32, 397, and in narratives pertaining to the *Globe* mutiny, no specifications for the ship exist. The ones given here would be typical for a whaleship of the first half of the 19th century. See Henry Hall, *Ship-Building Industry of the United States*, pp. 25–26, where specifications for a vessel about the *Globe*'s size are spelled out in detail.

7. "The class of ships built immediately after the last war were about 300 tons burthen." Obed Macy, *History of Nantucket*, p. 215. Macy goes on to discuss construction and cost. (See next note.)

8. Authorities differ as to exact costs of vessels in this era. Henry Hall, *Ship-Building Industry*, p. 87, says costs in

1825 were $75–$80 per ton, but on p. 105 cites a Maine firm in 1825 building for $45 per ton. Briggs, *History of Shipbuilding*, p. 234, cites a cost of just over $10,000 for a vessel of about the *Globe*'s size built in 1819.

9. William Sutherland wrote the first textbook for English shipwrights in 1711, but this book only formalized methods Phineas Pett, master builder for the Royal Navy, might have used in 1600. See Abell, *The Shipwright's Trade*, pp. 26–107, for a discussion of the slow evolution of shipbuilding technology.

10. Briggs, *History of Shipbuilding*, p. 231, gives the *Globe*'s measurements and some of her history.

11. Ibid., p. 69. See also the works of shipbuilding historian Dana Story, particularly *Frame Up!* and *The Shipbuilders of Essex*, for details of life in the yards during the days of wooden ships and for the best descriptions, anywhere, of wooden shipbuilding methods.

12. Henry Hall, *Ship-Building Industry*, p. 102.

13. According to Briggs, the height of the

lower deck space on the *Globe* in the forecastle and blubber room was five feet, six inches. This is not to imply that deck height was strictly governed by man height. However, for practical reasons, men of average height had to be able to work there. She probably had a raised quarterdeck, which would have afforded six feet of headroom in the cabin. Briggs, *History of Shipbuilding*, p. 231.

14. These libations seem to have been reckoned as a part of the shipbuilding cost. For example, after quoting a cost of $25 for a capstan, Briggs cites the $1.14 spent on twelve gallons of cider at the launching of the *Peruvian*, another whaleship belonging to Christopher Mitchell and Sons. Briggs, *History of Shipbuilding*, p. 234.

15. See Kemp, *Oxford Companion to Ships*, pp. 703–07, for a concise discussion of sailing-ship rigs.

16. The complexities are such that, for three centuries at least, entire books have been devoted to them. Perhaps the best modern work is Harland's *Seamanship in the Age of Sail*.

17. A *block* was simply a wheel or pulley set in a protective block of wood.

Tackle (pronounced TAY-kul) was the collective term given to a system of blocks and lines arranged in such a way as to increase their mechanical advantage. Ship modeler and marine historian Erik Ronnberg adds the following: "Properly defined (by a sailor, of course) a *tackle* consists of one or more *blocks* and the *fall* (rope) rove through them. 'Block and tackle' is one of those landlubber's terms made valid through mindless repetition."

18. Starbuck, *History of the American Whale Fishery*, p. 219.

Chapter Five: Comstock and Worth

1. William Comstock refers to two Nantucket shipmates aboard the *Foster*, John Lincoln and John Cotton, with whom Samuel felt intimate enough to propose that they gang up and flog one of the officers. Lincoln and Cotton both demurred, and Comstock "gave up the project." Lincoln and Cotton, following the expected Nantucket career path, became whaling captains. Comstock, *Life of Samuel Comstock*, p. 42. For Lincoln and Cotton, see listings in

Lund, *Whaling Masters.*

2. In 1824, Glover wrote a long poem that was in part favorable to Comstock's memory. It was called "The Young Mutineer," and it was published in the *Nantucket Inquirer.*
3. Comstock, *Voyage to the Pacific,* p. 69.
4. Ibid., p. 68.
5. Crevecoeur, *Letters from an American Farmer,* p. 178.
6. Ibid., p. 179.
7. This and other unattributed quotes in chapter 5 are from Comstock, *Life of Samuel Comstock.*
8. The logbook for this trip, the only one surviving for the *Globe,* is at the Nantucket Historical Association.
9. For example, the crew of the *Maria* finished her construction and did all the rigging work on her in the fall of 1822. A manuscript account of this by George Washington Gardner Jr. is at the Nantucket Historical Association.
10. When stowing, and presumably when off-loading oil, the first mate customarily had command of affairs on deck, and his authority could extend into the hold as well. See p. 286 of the article "The Whale Fishery" by

Clarke and Brown in Section V of Goode's *The Fisheries and Fishery Industries of the United States.*

11. Ibid., p. 239. The *Globe* probably dated from the days of three-tier ships. This designation referred to the layers of casks stacked in the hold.

12. See chapter 4, note 3 indicating Coffin's interest in Cambridge insurance companies. The information about Worth is taken from a recollection of William Easton published in the *Nantucket Inquirer.* "Captain Thomas Worth who commanded the Globe on her fourth voyage, commenced his sea-life in her, advancing, if I remember correctly, from seaman to second mate, then mate, then finally met his sad fate, when master on his fourth voyage. Capt. Worth came from a counting-room in Boston when he joined the Globe, and was a young man of rare accomplishments. Well do I remember his fine personal appearance and gentlemanly manners." This clipping is contained in collection 129, book 8 at the Nantucket Historical Association.

13. See Stackpole, *Sea-Hunters,* pp. 266, 303, 354, for a few highlights of

Gardner's career. According to his own statement, his sailing days covered a million miles and thirty-seven years during which he "was at home but 4 years and 8 months." Obed Macy, *History of Nantucket*, p. 214.

Chapter Six: The Brothers Comstock

1. "The work whalemen performed, the sea they encountered, the relationships they sustained, and the conflicts they endured, all took their toll. . . . The sea could be a destabilizing element. . . ." Creighton, *Rites and Passages*, p. 6. However, in Samuel's case there appears to be a morbid or pathological tinge to it.
2. This and other unattributed quotations in chapter 6 are from Comstock, *Life of Samuel Comstock*.
3. Admiral George Emery helped me develop this alternate interpretation. See also McKee, *A Gentlemanly and Honorable Profession*, and Edward Billingsley, *In Defense of Neutral Rights*, both of which deal with the patriot struggles and America's response to them.
4. See Leach and Gow, *Quaker Nantucket*, p. 159.

5. So William reports in his novel, where the vessel is called the *Captain S.* Judging from clues sprinkled liberally through his whaling novel, William actually sailed on the *Governor Strong*, which departed for the Pacific in 1820. William includes a lengthy passage about the death of one of the mates, a man named Swain. Regarding the 1820 voyage of the *Governor Strong*, historian Alexander Starbuck notes, "Benjamin Swain, mate, died on the voyage." Starbuck, *History of the American Whale Fishery*, pp. 232–33. It returned in January 1822, in time for William to have met his brother in New York. However, as he related events in his biography of Samuel, William is either distorting the chronology of his whaling career or he left on a second voyage in 1822.

6. Historian Erik Ronnberg suggests that the lee side of a clew line would be hard hauling, and that the bitter end of a jib downhaul would be a lubber's position, indicating loss of status.

Chapter Seven: Christopher Mitchell & Co.

1. By 1822, when the *Globe* returned

from her third voyage, her ownership had gone over from P & C Mitchell to Christopher Mitchell & Co. This probably indicates nothing more than a decision by Paul and his brother Christopher to each manage his own firm. The actual investors in the vessel — the Mitchells, and their various sons and partners — could have remained essentially the same. However, under Christopher's regime, a few changes were in store. Starbuck, *History of the American Whale Fishery*, p. 232. Briggs, *History of Shipbuilding*, p. 231.

2. This is a measure of carrying capacity, calculated from the dimensions of the vessel, rather than gross weight.

3. G. W. Gardner Jr.'s manuscript at the Nantucket Historical Association confirms the ownership change. "Christopher Mitchell & Co. the owners of the ship Globe that had been three voyages in the Pacific. . . ."

4. Account by William Easton at the Nantucket Historical Association, cited in chapter 5, note 12. "The Ship Globe . . . sailed from this port in December 1815, in command of

Capt. George Washington Gardner, who made three voyages in her . . . all accomplished with the same cables and anchors. This was noteworthy, from the fact that rarely ships, if ever, performed these Pacific Ocean voyages without having at least one new cable."

5. Slush was a mixture of linseed oil and tallow soap used to grease the masts. This composite of typical refitting tasks is assembled from Leavitt, *The Charles W. Morgan*, p. 45; Clarke and Brown in Goode, *The Whale Fishery*, pp. 232–40; Henry Hall, *Ship-Building Industry*, pp. 22–28.

6. There is no conclusive evidence that the *Globe* carried spare whaleboats. If she did, they'd have been stored upside down on this same rack.

7. As another way of calculating these proportions, Erik Ronnberg notes that by the 1850s the cost of readying a new ship was reckoned as: $1/3$ for building hull; $1/3$ for finishing hull, rigging, and sails; and $1/3$ for whaling gear and provisions.

8. See the discussion on lays that follows in the text. The experienced Captain Gardner would likely have negotiated

a shorter lay than first-time captain Worth.

9. A document in the records of the Dukes County Probate Court shows that Worth had loaned money to the Kidder brothers. Such transactions between captain and foremast hands were occasionally made during a voyage, but could also have been done on shore prior to the ship's departure. Peter, incidentally, is described in the crew list as being a resident of Nantucket. He may have just returned from a whaling cruise when he signed aboard the *Globe*.

10. See Jenkins, *A History of the Whale Fisheries*, p. 63. Hohman speculates that the system of lays originated in the Dutch Greenland fishery during the 17th century. Hohman, *The American Whaleman*, p. 217.

11. See the discussion in Hohman, *The American Whaleman*, chapter 10.

12. So William asserts in his biography, p. 69.

Chapter Eight: Irene

1. Irene's copy went to the New Bedford Whaling Museum.

2. Unbeknownst to me, there was some

awareness of the matter in the early eighties. The text of Comstock's *Voyage* was "discovered" and reprinted in 1976 in *American Transcendentalist Quarterly*, no. 29. In a brief introduction, Joel Myerson points out that Melville specialist Howard P. Vincent was also aware of the Comstock brothers and their influence on Melville. In 1987, the Comstock books appeared in Bercaw's *Melville's Sources*.

Chapter Nine: William Lay

1. Cook and Jarrett were described as mulattos on the *Globe*'s crew list. Interestingly, neither is mentioned in the narratives of Comstock, or Lay and Hussey.
2. See chapter 7, note 9 for more on Kidder's Vineyard origins.
3. It's an odd name, which is handwritten in the crew list and in George Comstock's narrative. In both places its spelling is unclear. Stackpole calls him "Nahun McLurin." Hoyt calls him "Nahan Mahurin." The relative who transcribed Comstock's account calls him "Nahan Mawin." From an examination of the name's two

appearances on the crew list, I'd suggest "Nahum McLurin." However, none of these names appeared in the Connecticut census, probate records, or vital statistics for the 1800–1825 period at the New England Historical and Genealogical Society.

4. Sansom, "A Description of Nantucket" in *The PortFolio*, vol. 5, (1811).

5. Obed Macy, *History of Nantucket*, p. 214.

6. Hohman, *The American Whaleman*, p. 98, cites a figure between $60 and $100. Creighton, *Rites and Passages*, p. 36, suggests $75 in a slightly later era.

7. A note on the back of the *Globe*'s crew list reads, "This may certify that Nahum McLurin mentioned in the within list was taken out of the within named ship by order of law previous to his leaving this port and Constant Lewis is entered in his stead."

Chapter Ten: Comstock's Cold Eye

1. Stuart Frank, curator of the Kendall Whaling Museum, speculates the portrait may have been the work of prolific Vineyard artist Frederick Mayhew, and that Mayhew could

have been related to Worth's wife, Hannah (Mayhew).

2. Ashley gives a good rendition of the traditional captain's speech in *Yankee Whaler*, p. 5.

3. Ibid., p. 6.

4. See Browne, *Etchings of a Whaling Cruise*, p. 204. Browne lists nine basic messages that his ship could send to its boats using signal pennants, sails, or a combination of the two. Thus, "*Whales on the lee bow* — Lee clew up" or "*Boat stove* — Colors at the fore and mizzen." These signals were particular to each vessel, ensuring that other ships could not intercept them.

5. Or the topgallants, if the *Globe* was whaling. Generally, on the whaling grounds, she'd have sent her royal yards down. In either case, this was before the era of mast hoops. At the tops of the royal or topgallant mastheads, two short timbers known as *crosstrees* were installed, projecting horizontally across the topgallant yard. The watch was expected to stand on these for his two-hour stint. Ninety feet above the deck, the ship's rocking motion would be magnified

enormously. A novice's first watch aloft must have been memorable. Melville compared it to standing on the horns of a bull, but added that even the greenest hands soon grew comfortable on the lofty perch.

6. See, for example, Philbrick, *Away Off Shore*, p. 162.
7. These logs are bound in a single volume, which is kept at the Nantucket Historical Association.
8. See chapter 6, note 5 regarding William's somewhat confused chronology.

Chapter Eleven: The Greenhand's Education

1. More precisely, Erik Ronnberg notes, "A common logbook passage, 'all hands employed setting up the rigging' refers to the continuous process of taking up slack in shrouds, stays and backstays, cutting the seizings, stretching the stays, backstays and shroud lanyards with 'setting-up tackles,' then re-seizing them."
2. Ashley, *The Yankee Whaler*, p. 59.
3. The first centerboard in a whaleboat appeared sometime in the 1850s, but they did not come into general use

until the 1870s. Without this addition, there was no way of preventing leeway — of holding the boat's course — while under sail. So the use of sails was limited prior to the introduction of the centerboard.

4. If mast and sail had been aboard, they would have run the length of the boat and extended over the stern. However, historian Mary K. Bercaw points out that masts and sails were not routinely carried in the early 1820s.

5. Some time after the *Globe*'s era, hemp was replaced by manila, which was less liable to rot when wet.

6. Manuscript by Gardner Jr., circa 1876. At the Nantucket Historical Association.

Chapter Twelve: First Trial

1. "Nothing can be more bewildering to a youth, whose imagination naturally magnifies all dangers of the deep, than to be roused up in the dead of night, when the ocean is lashed into a fury by a stiff gale, the vessel pitching and laboring, and the officers yelling at the men as if endeavoring to drown the roaring of the elements." Browne, *Etchings of a Whaling Cruise*, p. 24.

2. Hohman, *The American Whaleman*, p. 151.

3. Stewart in his *Visit to the South Seas*, vol. 2, 214–16, observes that American whalemen usually refitted in Hawaii in April and May, and then proceeded to Japan, where they remained until October.

4. Hemp rope is *wormed* by placing thin line into the lay of the rope; *parceled* by wrapping it in strips of canvas; *served* by wrapping the canvas with tarred thin line. This would have been accomplished by the riggers in port, but always needed redoing on damaged rigging.

5. Although the marine chronometer became a reality in the middle of the 18th century, it remained an expensive, "high-tech" appliance. Speaking of navigation in the late 1700s, Stackpole says, "Chronometer and sextant were used only by the circumnavigators." Stackpole, *Sea-Hunters*, p. 155. Things were essentially the same thirty years later. In fact, some whaling captains were incapable of the complex calculations involved in taking lunars, and relied solely on dead reckoning to

ascertain their longitude.

6. See Chapter 22, in which the fourteen-second glass is used. See also the discussion of the mathematics involved, in Kemp's *Oxford Companion*, pp. 492–94.

7. It would be a few more years before decorative scrimshaw put in its appearance among American whalemen. Stuart Frank of the Kendall Whaling Museum suggests 1824 as the earliest date. See also Frank, *Dictionary of Scrimshaw Artists*, pp. 5–9 and Basseches and Frank's monograph, *Edward Burdett, 1805–1833. America's First Scrimshaw Artist.*

8. Harland, *Seamanship*, pp. 10–11, 189. There were thirty-two points on the seaman's compass, each representing a change of 11.25 degrees, to be gabbled off.

9. Olmstead, *Incidents of a Whaling Voyage*, p. 71, talks of keelhauling when crossing the line. J. Ross Browne, a colorful writer who wouldn't miss a chance to report such a ceremony, makes no mention of any special rituals when his ship sailed over the equator. Busch, *Whaling Will Never Do For Me*, p. 160, speaks of the

whaleman's "disdain for the standard maritime ceremony held upon 'crossing the line.' "

Chapter Thirteen: The First Whale
1. It was not until the 1850s that a truly improved harpoon came into general use.
2. On this voyage, the cook was not one of the shipkeepers. They may have been shorthanded, or the cook may have been too good a man to keep out of the boats. This and the "Mr. Oisten" quotation are from the *Globe* logbook at the Nantucket Historical Association.
3. Goode, *Fisheries and Fishery Industries of the United States*, section V, pp. 272–73.

Chapter Fourteen: Cutting In
1. Obed Macy, *History of Nantucket*, p. 223.
2. By the 1860s, this jury-rig had been superseded by a structure called the *cutting stage*, which was a prefabricated platform with better footing and primitive railings and guards. It was safer to work from, though lives were still lost while cutting-in.

3. Melville, *Moby-Dick*, p. 1,117.
4. Historians Mary K. Bercaw Edwards and Glenn Grasso say Melville is the only source for this, but it seems logical enough to secure a man to the ship. I am assuming that it was employed on other vessels besides the *Pequod*.
5. Starbuck, *History of the American Whale Fishery*, p. 301, claims that in 1836 the *Wade* of Dartmouth, Massachusetts, returned with fifty barrels of the precious stuff, but this is almost certainly an error. See Hohman, *The American Whaleman*, p. 4, and Creighton, *Rites and Passages*, p. 47.
6. "Cutting in, trying out and clearing up the decks, occupied us for the next six days. . . . Working incessantly in oil, which penetrated to the skin, and kept us in a most uncomfortable condition, besides being continually saturated with salt water, produced a very disagreeable effect upon those who were not accustomed to such things, by chafing the skin, and causing painful tumors to break out over the whole body." Browne, *Etchings of a Whaling Cruise*, p. 135.
7. David Porter, the hero of Pacific

whalemen in the War of 1812, had rounded the Horn in about the same season. His experience may stand for theirs: "Although we deemed ourselves more fortunate than other navigators had been, in getting around Cape Horn . . . yet we had not been without our share of hardships. The weather had for some days been piercing cold; this, with the almost constant rains and hails, and the water shipped from heavy seas, and from leaks, kept the vessel very uncomfortable, and the clothes of the officers and crew very uncomfortably wet. The extremities of those who had formerly been affected by the frost, became excessively troublesome to them . . . and as to refreshments of any kind, they were entirely out of the question . . . deceitful intervals of moderate weather would for a moment encourage us to make sail; when, in a few minutes afterwards, blasts, accompanied by rain and hail, would threaten destruction to our sails and spars."

8. Lay says, "We saw whales only once before we reached the Sandwich Islands, which we made on the first of

May early in the morning." George Comstock's manuscript account reads, "We saw whales but once until we reached the Sandwich Islands, which we made on the first of May 1823 early in the morning. . . ." Both narratives so resemble each other that the question arises as to which writer was a source for the other. Logically, George's account would have been written for the owners just after his return and should have preceded Lay's. As discussed in a later chapter, Lay and Hussey probably employed a ghostwriter, and this individual may have consulted George's account.

Chapter Fifteen: The Golden Age
1. Britton Cooper Busch makes the following interesting statement: "Scholars seem now to be agreed that the negative qualities of nineteenth-century Western impact have been over exaggerated, in the sense that small Pacific societies were not simply overwhelmed, despite the impact of both diseases and technologies, but rather adapted that impact in ways that yet preserved the essentials of their own societies." Busch, *Whaling*

Will Never Do For Me, p. 101.

2. In an article that was probably an accurate representation of 19th-century thinking, the *Nantucket Inquirer* for May 9, 1822, said, "The Sandwich Islands are now becoming a place of great commerce, and the Natives making rapid strides towards civilization. . . . There are now residing amongst them several of the Missionary Society from the United States . . . the moral and religious precepts delivered by the Rev. Mr. Bingham . . . is daily increasing amongst those children of nature. . . . The Natives themselves are now the owners of ten square rigged vessels, none less than 120 tons, besides a number of Schooners and Sloops which they keep constantly going from Island to Island with Sandal Wood, provisions, &c. . . . On the south side of the Island of Woahoo . . . can be obtained refreshments of every kind, and a ship can be repaired if needed — for this last year it has been a resort for all the Whale ships."

3. Rufus Anderson says, "Kamehameha was a remarkable man . . . he was wounded by one of the guns fired at

the time Captain Cook was killed." Anderson, *The Hawaiian Islands: Their Origin, Progress, and Condition*, p. 36.

4. On the whaleship *Montpelier* in 1844, for example, four men refused duty "because they could not have women aboard the ship." *Montpelier* log, private collection.

5. Smith's quote is taken from Consul Hogan's deposition, as were George's and Stephen Kidder's, below.

6. Busch, *Whaling Will Never Do For Me*, pp. 19–21. See also, Creighton, *Rites and Passages*, pp. 92–94.

7. Hohman, *The American Whaleman*, p. 62.

8. Ibid., pp. 62–64. The author examines thirty-six voyages made by thirteen vessels during the period 1839–1879. He finds 326 desertions.

9. Busch, *Whaling Will Never Do For Me*, p. 104. Ashley, *Yankee Whaler*, p. 104.

10. Modern authors, such as Hoyt and Stackpole, could only guess at the date of the *Globe*'s return. However, Pauline N. King's recently published *Journals of Stephen Reynolds* establishes the dates of the *Globe*'s arrival and departure as November 7 and

December 9. She put in and departed with the whaleship *Lyra*, indicating that they were whaling together. Reynolds was a merchant in Oahu.

Chapter Sixteen: Abandoned Wretches and Cruel Beings

1. "The chiefs, beguiled by the ease of signing notes to be paid at some future time, soon found themselves entangled in a wilderness of debts." Kuykendall, *The Hawaiian Kingdom*, vol. 1, 88–92. Also, Morison's colorful description, "It so happened that the panic of 1819, making it difficult to procure specie for China, coincided with a new reign in the Sandwich Islands, which took the lid off the sandalwood traffic. Kamehameha I had conserved this important natural resource, so much in demand at Canton. But . . . (Kamehameha II) a weak-minded and dissolute prince, cheerfully stripped his royal domain in order to gratify tastes which the Boston traders stimulated. They sold him on credit rum and brandy, gin and champagne, carriages and harnesses, clothes and furniture, boats and vessels; until he had tonnage and

liquor enough for an old-time yacht club cruise." Morison, *Maritime History*, p. 262.

2. Stackpole says he was a "Cape Cod Indian," but does not give a source for this information. Stackpole, *The Mutiny on the Whaleship Globe*, p. 10.

3. Gilbert Smith, in his deposition before Consul Hogan, says one of the men was discharged. Historian Edouard Stackpole says Cleveland was discharged by Captain Worth "because of illness." Ibid., p. 13.

4. Stackpole and Hoyt both favor this theory. William says that Samuel bullied the men he found sleeping on his watch, giving some credibility to the idea. Certainly it lends itself to the notion that Samuel's crime was premeditated.

5. Busch records one case in 1883 in which the master of the steam whaler *Lucretia* prevented a mutiny by shooting the ringleader to death. In another instance, the cook aboard the whaleship *Milton* in 1853 was shot for insubordination. Busch, *Whaling Will Never Do For Me*, p. 25.

6. Work stoppages, a less extreme form of rebellion, were far more common.

However, they were often less effective than mutinies. "Whalemen with experience . . . could not but realize that the result of disobedience . . . was very probably going to be physical punishment or imprisonment or both. . . . That such incidents continued to occur . . . is explained by the continued sense of injustice or sheer physical danger. . . . Always there was the hope that 'this time' the result would be victory (26 percent) or at worst a standoff (6.5 percent)." Ibid., pp. 51–61.

7. "Once well away from the vessel, a deserter looked for safety in numbers, hence the tendency to make for some sailortown, if such was available, where the law was reluctant to follow. . . . If shoreside authorities were hostile to deserters, and local residents more attracted by tangible rewards for runaways . . . the officers and men of another vessel might, in the proper circumstances, offer refuge." Ibid., p. 98. See Paulding, *Journal of a Cruise*, pp. 225–26.

8. It sometimes happened that a crewman would ship for a single passage only.

9. Alternately, these may have been the only men available. However, when the *Dolphin* put in at Honolulu two years later, the port was thick with deserters. See Paulding, *Journal of a Cruise*, pp. 225–26.
10. Hohman, *The American Whaleman*, p. 111. Legislation passed in 1803 required seamen to be discharged with three months' pay. Busch, *Whaling Will Never Do For Me*, p. 24. However, there are cases of unscrupulous captains simply abandoning sick men at whatever port was handy, or of captains discharging men to avoid paying out their share of the profits at the end of the voyage.
11. In his whaling novel, William Comstock writes about such a transfer, though in that case the captain of his ship came up a little short. The new crewman was discovered to have a glass eye. Comstock, *Voyage to the Pacific*, p. 32.

Chapter Seventeen: Comstock's Creatures
1. King, *Journals*, p. 8.
2. Maury, *Explanations and Sailing Directions*, plate 18.
3. Creighton, *Rites and Passages*, pp. 87–

93, discusses pressures on captains and their reactions.

4. Thomas's deposition before Consul Michael Hogan, Valparaiso, June 1824.

5. Comstock, *Life of Samuel Comstock*, p. 76.

Chapter Nineteen: The Loose Hatchet

1. William Comstock provides an alternate reading. He suggests Samuel kept bullying George to keep the ship more before the wind than was necessary. If she'd been sailing east by north, this might force her to a more easterly course. Her path would diverge from the *Lyra*'s, and the distance between them would increase. This, William points out, was just what Samuel would have wanted.

2. Thomas G. Lytle states that only officers were allowed to use boarding knives. However, Stuart Frank points out that it was the dangerous *task* of boarding blubber that was reserved for officers, rather than the *tool*. Lytle, *Harpoons and Other Whalecraft*, p. 143.

3. Blackburn, *Overlook Illustrated Dictionary*, p. 335, defines the waist of a

ship as "the middle part of a ship, usually that part of the deck between the first mast and the main mast." *The Visual Encyclopedia of Nautical Terms Under Sail*, p. 01.01, defines it as "that part of the ship between the quarterdeck and the forecastle."
4. In his manuscript account, George claims, "Then he went in the waist, I believe, for I see no more of him until he had committed the murder." It is difficult to imagine him not seeing the murderers descend the companionway, less difficult to imagine that he was too terrified to raise an alarm or stop the attack.
5. From William Lay's account. Lay, *A Narrative of the Mutiny*, p. 13.

Chapter Twenty: The Bloody Hand
1. Testimony of Gilbert Smith in the deposition taken by American consul Michael Hogan, Valparaiso, 1824.
2. Testimony of Joseph Thomas in the deposition taken by Consul Hogan, Valparaiso. All the depositions Hogan took relating to the *Globe* mutiny are in the American consular records at the National Archives in Washington, D.C.

Chapter Twenty-Two: The Fourteen-Second Glass

1. Kemp, *Oxford Companion,* pp. 675–76.

2. Lay and William Comstock say there were two men on the jury. George says four men. Regardless of their number, they would have been men aligned with Comstock. Gilbert Smith told Consul Hogan that Rowland Coffin was a juror. Several of the survivors accused Rowland Coffin of currying favor with the mutineers. (This will be discussed in a later chapter.) Mutineers Oliver and Lilliston would have been likely candidates for a four-man panel, joined perhaps by Joseph Thomas. Thomas's bad behavior and contact with the mutineers made him suspect by the crew of being sympathetic to Comstock.

3. Both George Comstock and William Lay write that Samuel and Silas Payne had decided to do away with the black steward.

4. Francis Allyn Olmstead, for example, paints a vivid portrait of his whaleship's cook who, "sustains a relation to the ship similar to that of the jester.

. . . He can sing a song, play upon the fiddle, dance various jigs . . . and roll up the white of his eye — all in the genuine negro style." Olmstead, *Incidents of a Whaling Voyage*, pp. 45–46.

5. "Stu'ns'ls" were rigged on the outsides of the regular square sails to obtain extra speed when the weather was fine and the wind was blowing from behind. The boom was simply a horizontal extension to the yard, allowing the studdingsail to be attached. See the place marked ✠ on the "*Globe*'s Sail and Rigging Plan" in the photo insert.

6. There is a fourteen-second glass on display in the East India Hall at the Peabody Essex Museum. Fourteen and twenty-eight seconds seem to have been common units of measurement for determining the ship's speed. See the discussion on p. 493 of Kemp's *Oxford Companion*.

Chapter Twenty-Three: Lord Mulgraves Range

1. They were so named by a British captain in honor of an admiral of the Royal Navy. As the brief history in coming pages will suggest, the nomen-

clature of these islands is confused, consisting of at least five layers — Spanish, Portuguese, English, Russian, and American — and further complicated by multiple discoverers and by chartmakers, who in their attempts to collate the sketchy information of early explorers, sometimes duplicated names or placed islands where none should be. This confusion has carried into our century, when the Gilberts are referred to by their native name, Tungaru, and are now part of the larger Republic of Kiribati (itself a corruption of the British colonial "Gilberts").

2. This peculiar spelling appears on Arrowsmith's "Chart of the South Pacific," which had been published in London about 1820. During the time of Wilkes's expedition, 1838–1842, they were still called the Kingsmill Group. George uses the same nomenclature in his narrative, suggesting the Arrowsmith chart or a variant was its source.

3. The islands are nowhere near on a straight line between Australia and China, but Samuel Eliot Morison explains the navigational complexities

of this leg of their voyage. "This was probably the first time that anyone had attempted to sail from Australia to China. It may seem strange that the two captains should make such a wide sweep to the eastward as to encounter the Marshalls. But the passage through Torres Strait was one that baffled even Cook . . . our captains probably figured on making a good easting in the westerly winds of south latitudes, in order to enjoy a fair slant in the northeast trades to Canton." Morison, "Historical Notes on the Gilbert and Marshall Islands," *American Neptune,* vol. 4, 93–94.

4. Kotzebue, *A Voyage of Discovery,* vol. 2, 7–8. This was on the island of Mejit, which Kotzebue called Christmas Island. It was a few islands up the chain from Mili Atoll in the Marshalls, the *Globe*'s ultimate landing place.

5. George's narrative has the *Globe* landing at Knox (Knoy) Island. Arrowsmith's chart places this at the northern end of the Gilberts, about where Tarawa ought to be. Morison suggests Knox Island was actually Makin Island. Morison, "Historical

Notes on the Gilbert and Marshall Islands," pp. 94–95.

6. The inhabitants of the Gilberts and Marshalls were renowned for their large seagoing outrigger canoes. Early accounts speak admiringly of these craft and of the skill of their owners in navigating long distances across the open Pacific. The Russian artist Choris included several striking lithographs of them in his *Voyage pittoresque autour du monde*, Paris, 1822. The whaleboats of the *Globe* would have been no match for them. Clearly the "canoe" referred to above was a smaller, cruder vessel.

7. George says this delay put the mutineers in an "ugly and malicious" mood.

8. In their published reports, Kotzebue and his colleagues made reference to the comeliness of Micronesian women, but they did not mention women being used as articles of trade. By the 1840s, however, when Charles Wilkes and the U.S. Exploring Expedition stopped at an island just south of Tarawa, their ship was approached by the usual swarm of native craft. "They brought off nothing except a

few cocoanuts; but the object of their errand was not to be misunderstood, for in each canoe there was a woman, which I think does not speak much in the praise of the whalers or other ships that frequent this cruising-ground." Wilkes, *Narrative*, vol. 5, 64–65. Shortly before this incident, a seaman from the expedition had been ambushed and killed by natives. While there were many islands, and many varied responses to early contacts with white men, it is tempting to recall William Lay's prophecy some sixteen years earlier and to think of Wilkes's comment about the use of women as articles of trade. The Globe was not setting a good precedent in these waters, and things would only get worse.

9. Stephen Kidder's account, as recorded by Obed Macy in a manuscript at the Nantucket Historical Association. This independent account is of interest primarily because it validates information given by other witnesses.

Chapter Twenty-Four: The Man of Blood

1. Edouard Stackpole in *The Sea-Hunters*, p. 431, says this dramatic

phrase occurred in a letter Nathan Comstock wrote to relatives in Nantucket. The quotation was repeated by later writers, including Snow and Hoyt.

2. "The reader will doubtless inquire, what was the object of our hero, in killing the Captain and officers of the Globe, and taking the ship to the Mulgrave Islands, to there be dismantled and destroyed. In reply, I can only say, that it had long been a favorite scheme of his, to establish himself on one of the Pacific Isles — to gain sufficient influence over the natives to induce them to elect him their king — and to live a daring and dreaded outlaw in his adopted clime." Comstock, *Life of Samuel Comstock*, p. 99.

3. In his earlier days, Dr. McGee had been a scuba-diving special-weapons expert on an elite deep-penetration unit in the U.S. Marines. Later, he became director of law enforcement and forensic services at Sheppard Pratt Hospital in Baltimore, Maryland. By the time I knew him, he was, among other things, a member of the hostage negotiation teams of the Del-

aware and Maryland State Police, and special consultant to the FBI's Critical Incident Response Group. His profile of the Unabomber was the basis of a search warrant allowing officials into Kaczynski's shack, and, as a result of his 1983 stint as team psychologist for the Baltimore Orioles, he is the only man in his profession to have earned a World Series ring.

4. For a detailed description of the dysfunctions discussed, see *Diagnostic and Statistical Manual of Mental Disorders. Fourth Edition, Text Revision,* known in the trade as *DSM-IV-TR.* This is a 1,000-page tome, painstakingly assembled over the years by mental-health experts, describing the hundreds of ways the human psyche can run off its rails. Personality disorders are covered on pp. 685–729. What Dr. McGee called Mixed Personality Disorder is referred to on p. 729 as "301.9 Personality Disorder Not Otherwise Specified." Hypomania is described on pp. 365–368.

Chapter Twenty-Five: The Great Escape
1. For example, Ian F. McLaren's bibli-

ography, *Laperouse in the Pacific,* lists over 1,300 books and articles. Beddie's *Bibliography of Captain James Cook* contains over 4,800 items. The Russian explorer Kotzebue wrote a drama based on Lapérouse's adventures, and it was quickly translated into Dutch, English, French, and Italian. Philosophers and politicians seized on the utopian potential of the South Pacific, and the French government greatly enhanced its knowledge of that part of the world by sending out ships in search of their lost explorer.

2. Peter Dillon's book was entitled *Narrative and Successful Result of a Voyage in the South Seas to Ascertain the Actual Fate of Lapérouse's Expedition.*

3. Ibid., vol. 2, 104.

4. Some of these men may already have been stationed as watchmen aboard the *Globe.*

5. According to Smith's deposition before Consul Hogan, Smith and his men had fallen in with a Chilean vessel about thirty miles south of the harbor, and some of the Chilean sailors had come aboard as pilots, helping them navigate into Valparaiso.

Benjamin Morrell said Valparaiso was a difficult port to enter. Ships should approach from the south, as the *Globe* had done, since there were prevailing southerly winds most of the year and a point of land at the southern end of the harbor that had to be rounded close to shore.

6. Gilbert Smith, in his deposition to Consul Hogan said, "When night came on we commenced our preparations for getting away with the ship, every man on board exerting himself to the utmost of his ability to get the ship ready, the deck being entirely covered with sails, rigging, casks, iron hooks, the charts & two compasses were also in the boats, the charts & compasses were taken on board." William Lay gives a slightly different story of the "charts & compasses," saying that Payne had ordered the ship's two binnacle compasses brought ashore, but that Smith hid one aboard and substituted the captain's hanging compass for it in the shore-bound shipment. However, whether there was one compass or two, compasses and charts were the only navigational equipment left

aboard the *Globe* that evening. Smith is explicit about the fact that they had no quadrant.

7. Stackpole, *The Mutiny on the Whaleship* Globe, p. 35.

8. Not being knowledgeable about the fine points of handling a square-rigged sailing ship, I thought I would get some expert opinion as to how those six (or five) young men could have managed it. To this end I sent e-mail inquiries to Mary K. Bercaw Edwards and Glenn Grasso. Mary K., as she is known to her colleagues, is a Ph.D. with a shoal of books and articles to her credit, and more than 50,000 miles under sail at sea. Glenn Grasso, a chantey singer and collector of chanteys, sailed on the Coast Guard square-rigger *Eagle* and is in the midst of completing a doctoral program. Both were customers of mine, so I could be certain that they were filling their minds with the highest quality maritime literature. More important (and seriously), both Mary K. and Glenn worked at Mystic Seaport in Connecticut, where their academic knowledge was supplemented by years of hands-on experi-

ence aboard the sailing ships preserved there. They also came highly recommended. Nathaniel Philbrick had consulted them both in the writing of his excellent *In the Heart of the Sea.* The analysis of how Smith and his men sailed back to Valparaiso is based in large part on their observations.

Chapter Twenty-Six: The Mysterious Thain
1. Morrell, *Four Voyages,* pp. 108–10.
2. Some small mystery lingers concerning this vessel. William Lay says, "The persons on board [of the *Globe*] were put in irons aboard a French frigate, there being no American man-of-war in port." According to Commodore Isaac Hull's biographer, there were five French naval vessels in the Pacific at that time, including the *Marie Thérèse,* a sixty-four-gun ship-of-the-line, but there is no specific mention of any of them being in Valparaiso. (Linda Maloney, *Captain from Connecticut,* p. 393.) Edouard Stackpole, in his account of the *Globe* mutiny, probably confused Dillon's search expedition in 1827 with his presence in Valparaiso as a trader in 1824, and imagines that Dillon was in

charge of the French frigate. (Stackpole, *Mutiny on the Whaleship Globe*, p. 36.) Still, he's on the right track. In his narrative, Peter Dillon said that his East India Company trading vessel, the *Calder*, "mounted 16 guns." This might make her seem close enough to a frigate to the person who transmitted this slightly wrong secondhand information to William Lay. It is quite possible that the men were confined aboard Dillon's vessel.

3. For the description of Hogan and Valparaiso, see Morrell's *A Narrative of Four Voyages*, pp. 109–11. Like his contemporary Peter Dillon, Morrell was a great spinner of yarns; in fact he's been called "the biggest liar in the Pacific." However, Morrell's words about Hogan come early in his book, before he got himself cranked up to maximum lying velocity. For the "biggest liar" attribution, see the citation for Morrell in de Braganza's bibliography of the *Hill Collection of Pacific Voyages*, pp. 203–04. De Braganza points out that Morrell's wife accompanied him on his fourth voyage in 1829, and wrote her own fascinating and early account of the Pacific.

4. The record of Hogan's questioning of Smith and his men survives on about fifty handwritten pages in the American Consular Records stored at the National Archives in Washington, D.C. It is likely that a secretary was present to record the questions and answers, and that at least two scriveners worked on making fair copies of the resulting lengthy document, since the surviving copy is transcribed in two different hands. The consul took his depositions in three groups of two men each. He first questioned Stephen Kidder and George Comstock on June 9 and 10, then Peter Kidder and Gilbert Smith on June 15, and finally Anthony Hansen and Joseph Thomas on June 30.

5. A letter from one John Haven, a crewman on the whaleship *Loper*, written from somewhere in the North Atlantic in February 1825, refers to the mutiny, which "was done cause they could not get enough victules to eat. . . ." Haven's letter is at the Nantucket Historical Association.

6. On October 21, the *Boston Commercial Gazette* ran a similar story. They'd gotten their information thirdhand,

from a Captain Macy of the merchant ship *Palladium*, which had just come in from Coquimbo, another port in South America. Macy had gotten his information from the captain of the *Belle*, whom this article identified as a man named Eddiston. The *Belle* was the ship that first brought the news to Boston. Next day, the *New Bedford Mercury* ran the same article, adding Captain Eddiston as the informant and incorrectly giving the location of the Mulgrave Islands as seven degrees south, rather than seven degrees north. The article was reprinted in the *Nantucket Inquirer* on October 25.

7. *Nantucket Inquirer*, November 15, 1824.

8. This incident had such a profoundly negative effect on the residents of Nantucket that their first historian, Obed Macy, who was there to receive the news in 1824 and who even recorded Stephen Kidder's testimony in his private journal, made no mention of the *Globe* mutiny in his *History of Nantucket* first published in 1835. Not even the *Essex* tragedy of 1820, or the painful and divisive bank robbery of 1795 was ignored in such a

fashion. (The Nantucket vessel *Essex* was sunk when rammed by a whale. Survivors resorted to cannibalism. The full story is told in Nathaniel Philbrick's *In the Heart of the Sea*. In 1795, $20,000 was stolen from Nantucket's first bank. It was almost certainly an inside job, and the accusations that followed in its wake divided the community for generations.)

9. Starbuck, *History of Nantucket*, p. 629.
10. Smaller journals, such as the *Gazette and Patriot*, a weekly paper in Haverhill, Massachusetts, ran the story in its entirety.
11. The circuit courts were abolished in 1911.
12. The indefatigable Edward Rowe Snow, author of dozens of books on New England's maritime history, wrote a treatment of the *Globe* affair framed by the trial of Joseph Thomas. He raises the interesting point that Thomas may have refused to take part in the mutiny because he was disabled — suffering the ill effects of his recent beating by Captain Worth. See Snow, *Piracy, Mutiny and Murder*, pp. 132–63.

Chapter Twenty-Seven: The Admiral

1. It went to the Kendall Whaling Museum in Sharon, Massachusetts, now united with the New Bedford Whaling Museum.

Chapter Twenty-Eight: Uncle Gorham

1. According to one Nantucket historian, "Mr. Jenks retired temporarily from the editorship with the last issue of 1824, William Coffin Jr. taking his place. In a few months, however, he was again at the helm." Starbuck, *History of Nantucket*, p. 629. This account also places William Coffin Jr., possible ghostwriter of Lay and Hussey's narrative, right in the middle of the breaking news story.

2. See Obed Macy's manuscript, "A Journal of the Most Remarkable Events Kept by Obed Macy. 7mo 1822," at the Nantucket Historical Association.

3. According to Edouard Stackpole, on November 19, Smith was sworn in by the Edgartown customs officer as the new captain of the *Globe*. Stackpole, *Mutiny on the Whaleship* Globe, p. 36. This is substantiated by a note on the back of her 1824 enrollment certifi-

cate stating that Gilbert Smith had been made master of the *Globe*, replacing James King, who'd sailed her back from Valparaiso. For Smith's subsequent career, see Chapter 37.

4. Whaling quickly became a sufficiently powerful economic engine that it generated capital for other phases of expansion. "It seems clear . . . that the capital generated by whaling financed the construction of railroads in Massachusetts, the textile industry in New Bedford and Fall River. . . ." Lofstrom, *Paita: Outpost of Empire*, p. 20.

5. Obed Macy, *History of Nantucket*, p. 228. Head matter, or spermaceti, processed in 1816 — 682 barrels; in 1824 — 11,875 barrels.

6. In the 20th century, his abilities would be documented by Stephen Vincent Benet, whose story "Daniel Webster and the Devil," has D. W. out-arguing Satan himself.

7. An accident of history aided Nantucket's cause. John Quincy Adams was about to succeed James Monroe in a presidential election as controversial in its time as the election of George W. Bush in ours. In the elec-

tion of 1824, Andrew Jackson had gained ninety-nine electoral votes, Adams eighty-four, William H. Crawford forty-one, and Henry Clay thirty-seven. Since there was no majority, the Constitution provided that the vote be sent to the House of Representatives. There, Clay swung his support to Adams's side, and on February 9, 1825, Adams won the presidency. He named Clay his secretary of state, much to the dismay of Jackson loyalists who immediately raised charges of bribery and corruption. To counter these, Adams took the unusual step of retaining the former president's appointees, rather than appointing his own supporters to key positions within the government. While this may have avoided the appearance of impropriety, it proved to be political suicide, since the new president was then crippled by not having his own people in place. Although the policy was unfortunate for Adams, it was a blessing for *Globe* rescue efforts because the competent Samuel Southard, Monroe's secretary of the navy, remained in that office under Adams's administration. This meant that, throughout the course of the rescue operation, there was a continuity of leadership. Leonard D. White, who surveyed naval administrators from 1801 to 1829, regarded Southard as "perhaps the ablest of the group." And Charles O. Paullin concluded that, "Of the nine naval secretaries of the period 1815–1842, Samuel L. Southard rendered the most efficient service." Coletta, *American Secretaries of the Navy*, vol. 1, 132.

Chapter Twenty-Nine: Uncle Isaac and Mad Jack
1. When James Biddle, commanding the sloop *Ontario*, entered the Pacific in the spring of 1818, he discovered that Valparaiso had been blockaded and that several American merchant vessels were being held inside the port. During a prolonged visit marked by bluff and intrigue, Biddle, acting more the diplomat than the sailor, succeeded in playing the Royalists (who depended on supplies delivered by Americans) against the Patriots (to whom the Americans were by nature sympathetic). He managed to secure

the release of the American ships and enforce the idea of our right to neutral trade, before continuing his cruise to the northwest coast. He also played a part in the release of Samuel Comstock's gunrunning ship, the *Beaver*. If not for Biddle, Comstock might have languished in jail for another couple of years and missed his chances to ship on the *Foster* and the *Globe*. Billingsley, *In Defense of Neutral Rights*, chapters 2–6.

2. Bestselling author Patrick O'Brian used Cochrane as a model for his hero Jack Aubrey. There can be little doubt that the commodores of the Pacific Squadron would rather have read about the heroic exploits of the fictional character than deal with the erratic actions of his real-life counterpart.

3. A contemporary letter from American Commercial Agent John C. Jones provides an excellent description of the conditions that had existed in Hawaii throughout the 1820s. "Thus annually about thirty thousand tons of American shipping visit these islands. . . . Owing to the frequent desertion of seamen at these islands . . . a

number of the most abandoned of that class of people have accumulated. . . . Such men as these, if not taken from here, will eventually, driven by hunger and want, commit depredations and piracies, without fear of an avenging arm." Paullin, *Diplomatic Negotiations*, p. 336. The letter was actually sent in 1826, but the conditions it described had been extant throughout the decade.

4. This and subsequent communications between Southard, Adams, and Hull are quoted from the series of documents at the National Archives, "Letters Sent by the Secretary of the Navy" and "Letter to the Secretary of the Navy."

5. Coincidentally, Herman Melville, on his way home from his adventures in the Marquesas, enlisted as an ordinary seaman on the *United States* at Honolulu during her second tour in the Pacific.

6. See Dean King, *A Sea of Words*, p. 248, and the definition of "master."

7. My favorite story about Percival was when, as an old man, he carried his own coffin with him on his last cruise around the world. Failing to die, he

took it home and used it as a watering trough in his front yard. Several such anecdotes are succinctly related in the article by Allan Westcott in *Dictionary of American Biography*, vol. 7, 461, and at greater length in Long's biography, *"Mad Jack,"* pp. 2–10.

8. The encounter is described in Bowen, *The Naval Monument*, p. 226.

9. Speaking of such men, Percival's biographer refers to "the impetuous and high handed David Porter, the sometimes devious plotter William Bainbridge, the snobbish and aloof James Biddle, the truly malevolent Jesse Duncan Ellitt [*sic*], and — yes — the hotheaded and self pitying John Percival." Long, *"Mad Jack,"* p. 32.

10. Ibid., pp. 40–41. Commodore Downes complained about the incident to the governor of Valparaiso who, in true revolutionary fashion, sentenced the offending soldier to be shot. Those were the days!

11. As Hull's biographer has it, his gut instinct was correct. "Considerations of favoritism aside, Percival was the logical choice for this command. He was an old and experienced officer —

at forty-eight a good deal older than most of his rank, since he had 'come up through the hawse hole' from sailing master; he had been on the coast previously as a lieutenant in the Macedonian, and he spoke Spanish." Maloney, *Captain from Connecticut*, p. 376.

12. Square-rigged on the foremast, fore-and-aft rigged on the main, and capable of converting to the standard schooner rig, these versatile craft are pictured and described in Chapelle, *The History of the American Sailing Navy*, pp. 324–30.

13. According to Chapelle, the great historian of sailing-ship technology, "There can be no doubt that the clipper schooners were faster than even the best of the modern seagoing yachts over a long course, but the clipper rig required a big crew to handle it." Ibid., p. 330.

14. This referred to the weight of the shot. Larger vessels might carry guns capable of firing shot weighing forty-two pounds or more.

15. Provisions required for the voyage included:
salt provisions [i.e., salt beef and

pork] 65 bbl [barrels]
flour 12 do. [ditto]
raisins 100 pounds
butter 300 pounds
cheese 734 pounds
beans 348 gallons
rice 212 gallons
vinegar 150 [gallons]
molasses 140 [gallons]
spirits 800 [gallons]
National Archives, Letters from Commanders, Hull to Southard.

16. Ibid.
17. Strong's correspondence is at the Cincinnati Historical Society.

Chapter Thirty: Augustus and Herman
1. Emery, *Historical Manuscripts*, p. 58.
2. Within five years, Hiram Paulding had published his memoir of the *Dolphin*'s cruise, and this perhaps inhibited Strong's attempts to publish his own account. Into his journal, as a sort of illustrated frontispiece, Strong has pasted a portrait of Hiram Paulding. There is little doubt that Strong considered First Lieutenant Paulding, who played a heroic role in this mission, as something of an idol.

3. The Spanish word for tortoise is *galapago.*
4. Stackpole reads this name as "Horner." But Strong's Vevay journal and the *Navy Record* leave little doubt that it is "Homer."
5. Porter describes it as a box, which was "nailed to a post, over which was a black sign, on which was painted *Hathaway's Postoffice.*" Porter, *Journal of a Cruise,* vol. 1, 128.
6. According to Paulding, they were kept on deck. The usual practice was to stow them in the hold, upside down. "The turtle . . . were very troublesome and offensive for a week, when they became quite domesticated and gave us not the slightest inconvenience." Paulding, *Journal of a Cruise,* p. 28.
7. Porter, *Journal of a Cruise,* vol. 2, 6.
8. Melville, *Typee,* p. 35.
9. Ibid., pp. 92–93.
10. So named by the equally Americo-centric Captain Porter who, in fact, renamed the island after President Madison. Porter, *Journal of a Cruise,* vol. 2, 80.
11. In his account, Paulding says they made this trip on the fourth. Paulding was reckoning by ship's time, which

ran from noon to noon. So, if they left on the morning of the fifth, it would still be the fourth by ship's time. Paulding, *Journal of a Cruise*, p. 61.

Chapter Thirty-One: The Ambassadors

1. C. H. Davis, in his biography of his father, who was a midshipman on the *Dolphin* during this cruise, points out that water was stored in wooden casks, and that it quickly became "foul and nasty" necessitating its frequent replenishment. *Life of Charles Henry Davis*, p. 42.

2. Long, *"Mad Jack,"* p. 61, identifies Duke of Clarence with Fakaofu, the southernmost island. But this is incorrect. See Wilkes, vol. 5, 10, and accompanying charts, "Groups in the Westernmost Part of the Pacific Ocean," which show Fakaofu to correspond with Bowditch's Island. See also Ward, *American Activities in the Central Pacific*, vol. 2, 304–18, and vol. 5, 269–74. Contacts with merchants in these islands aren't reported until the 1830s.

3. That phrase occurs in chapter 87, "The Grand Armada," in *Moby-Dick*.

The extended sentence of which it is the climax gives an excellent sense of the promise that the Pacific held. "Those narrow straits of Sunda divide Sumatra from Java; and standing midway in that vast rampart of islands, buttressed by that bold green promontory, known to seamen as Java Head; they not a little correspond to the central gateway opening into some vast walled empire: and considering the inexhaustible wealth of spices, and silks, and jewels, and gold, and ivory, with which the thousand islands of that oriental sea are enriched, it seems a significant provision of nature, that such treasures, by the very formation of the land, should at least bear the appearance, however ineffectual, of being guarded from the all-grasping western world." Melville, *Moby-Dick*, p. 1,199.

4. Samuel Eliot Morison's droll footnote is worth footnoting again here. He says that Commodore Byron was called Foul Weather Jack "by reason of his hard luck with the elements. His elder brother William was known as the 'wicked Lord,' and his son, John Byron, a notable debauchee, as

'mad Jack.' Lord Byron the poet was the son of 'mad Jack' and broke all family records." "Historical Notes on the Gilbert and Marshall Islands," *American Neptune*, vol. 4, 91.

Chapter Thirty-Two: Midshipman S.
 1. Strictly speaking, they would have headed northwest, or west northwest, since this was the orientation of Chirubon, most southeasterly of the islands on Mili Atoll. The *Dolphin*'s deck log has Lewis and Strong headed in this direction, but I have taken the word of Strong's firsthand account that they went "south" toward the lower part of the island.
 2. Mili is at the bottom of the eastern-most Ratak, or "sunrise," chain. Even in the 20th century these islands were considered sufficiently remote that they were used as test sites for America's atomic bombs.
 3. Paulding, in his account, refers to this land as stretching "to the south and west." Reference to modern maps makes this statement seem an error, as the land tends slightly north and west. But Paulding was referring to *magnetic* north rather than *true* north.

MULGRAVE ISLANDS.

1. The Landing place from the Ship Globe. 2. Mille Island, Lay's Residence. 3. Passage or the Small Boat, 4. The Straits. 5. The Ship Passage. 6. (Departure of the Dolphin). Hussey's Residence.

Old charts show compass deviation in those waters sufficient to make it seem that the islands tended slightly south as well as west. William Lay's chart, also based on magnetic north, demonstrates this effect.

4. The first search parties probably went ashore at the string of islets between Chirubon and Lukunor. Also, chronology of these events is a bit obscure because the *Dolphin*'s "day" ran from noon to noon. Thus, entries in her deck log for November 23 would also include events of the morning of November 24; whereas time in Strong's Vevay journal was marked by regular days.

5. There are detailed accounts of this

ticklish operation in both the Washington journal and the *Dolphin*'s deck log. It involved dropping anchors from boats at the stern of the vessel, and using the anchors to kedge their way off the bar.

Chapter Thirty-Three: Survivor

1. Based on Lay and Hussey's account, the island was probably Tokowa. The unattributed quotes describing Paulding's rescue of Lay are from Paulding's *Journal of a Cruise*.

2. Davis was a midshipman on the *Dolphin*. In the biography, *Life of Charles Henry Davis*, he claims to have been one of the twelve men in Paulding's boat. This is not substantiated by the *Dolphin*'s deck log, but it does not make Paulding's actions any less bold. As evidenced by an article by R. W. Meade in *Harper's Magazine* in 1879, Davis spent the rest of his days claiming he'd been there. Ironically, the only appearance Davis makes in the *Dolphin*'s deck log is at 9 P.M. on November 21, when he is reprimanded and sent below for lying down on watch.

3. Hussey was most likely on the island

of Jobenor.

Chapter Thirty-Four: The Massacre
1. The third boat had gone off with the *Globe*.

Chapter Thirty-Five: A Map of the End
1. The paper is watermarked 1813. The writing on the map was not done by the same hand that penned George Comstock's narrative, but it could be Gorham Coffin's writing, or one of the clerks at Christopher Mitchell & Sons who was present at Coffin's interrogation of George. Edouard Stackpole, the historian of Nantucket whaling, believed this map was made by George Comstock. After George wrote out his narrative of the mutiny, he drew a map of the island, made symbols on it to mark the main events, and explained the symbols to his interlocutor, who then transcribed the meanings of the symbols onto the map. Though unproven, this remains a reasonable assumption. In support of it are two pieces of circumstantial evidence: William Lay, William Comstock, and Stephen Kidder all spell Silas Payne's name "Payne."

George's narrative is the only one in which it is spelled "Paine," and so it is spelled on the map. In a letter dated December 22, 1824, Gorham Coffin, recounting events on the island, says that the mutineers "set about to build a hogstye." This is the only other reference to the "hog pen" mentioned in the legend to the map.

2. Tamar, wife of Jinwa, our translator on Mili, was the sister of Minister Lometo, in whose huts we stayed. Minister Lometo was a direct descendant of Langerine, the chief who had adopted Cyrus Hussey. Tamar, of course, owned all the land on which we stayed. True to Mr. Chutaro's description, she had a queenly fierceness about her.

3. They still are. One morning I woke to find a fifteen-foot aluminum boat pulled up on the beach in front of my hut. On inquiring I learned that the boat belonged to the husband of Jane, our cook. He'd brought some supplies over from Majuro for his wife — ninety miles of open Pacific in a fifteen-foot boat with a 15 h.p. Johnson on the transom. He'd left before midnight and gotten in at 6 A.M. No

lights, no charts, no loran, no GPS, no compass. Just the wind and waves and stars. "No problem."

4. Dirk H. R. Spennemann, "Mili Island, Mili Atoll, A Brief Overview of its WW II Sites," http://life.csu.edu.au/marshall/html/WWII/Mili.html.

5. The source is three photocopied pages of a typescript at the Alele Library on Majuro. The first page is headed, "*SURVEYS*, Mili Island." It bears the hand-printed name, Matsumoro, and appears to be one of the earliest postwar surveys of the island, predating more serious survey and reconstruction efforts in the 1960s and 1970s.

6. Rhodes was first elected governor of Ohio in 1962. If Chuji Chutaro had been born in the late 1930s, he would probably have been arriving in America at about this time.

Chapter Thirty-Six: The Diplomats
1. After their return from the Pacific, several of the *Dolphin*'s officers lived for a time in Flatbush, New York. In letters written in May 1827, Strong reports, "Mr. Paulding & Mr. Lewis

lodge with me . . . the former is a Lieutenant and showed me great attention during the cruise." And, two weeks later, "I still reside in this place persuing my studies, and relieving my mind mornings and evenings with an agreeable and improving walk, in company with my particular friend Mr. Paulding. . . ." (The Strong papers are at the Cincinnati Historical Society.) Michael Dyer of the Kendall Whaling Museum adds, "It is curious to note that Strong's account of the rescue of William Lay agrees precisely, almost word for word, [with] that given by the Dolphin's First Lieutenant, Hiram Paulding . . . suggesting that Strong's journal may have served as a partial basis for Paulding's narrative." Frank, *More Scrimshaw Artists*, p. 126.
2. Strong's journal.
3. Long, *"Mad Jack,"* p. 74.
4. Paullin, *Diplomatic Negotiations*, pp. 337–38.

Chapter Thirty-Seven: The Almanac
1. See the letter of April 27, 1827, at the Nantucket Historical Association, from Macy to Hussey Sr.

2. Letter at the Nantucket Historical Association. Historians Nathaniel Philbrick and Helen Winslow Chase each speculate this might have been William Coffin Jr. The same individual may have had a hand in Owen Chase's narrative of the sinking of the whaleship *Essex*, and Obed Macy's *History of Nantucket*. See Philbrick, *Away Off Shore*, p. 249.

3. This second edition is hair-raisingly rare, only one or two copies are known to exist. The one I consulted was in the Beinecke rare book library. It contains a map made by Lay twenty-four years after his rescue.

4. Information on the Kidders and Gilbert Smith from Stackpole, *Mutiny on the Whaleship* Globe, p. 60.

5. See chapter 28, note 3. By the *Globe*'s next whaling voyage in 1825, Reuben Swain had replaced Smith as captain. Starbuck, *History of the American Whale Fishery*, p. 254.

6. William Comstock's whaling novel, published in 1838, begins with a similar incident of a small boat upsetting in the harbor in Nantucket. If William was alluding to his brother's San Francisco adventure, George must

have left for the west well before the gold rush.

Bibliography

Diagnostic and Statistical Manual of Mental Disorders. Fourth Edition, Text Revision. Washington, D.C.: American Psychiatric Association, 2000.

Longworth's American Almanac, New York Register and City Directory. New York: Jona Olmstead, various years.

The Visual Encyclopedia of Nautical Terms Under Sail. New York: Crown Publishers, Inc., 1978.

Abell, Sir Westcott. *The Shipwright's Trade.* Jamaica, NY: Caravan Book Service, 1962.

Albion, Robert Greenhalgh. *The Rise of New York Port.* New York: Charles Scribner's Sons, 1939.

Anderson, Rufus. *The Hawaiian Islands: Their Origin, Progress, and Condition.* Boston: Gould and Lincoln, 1864.

Ashley, Clifford. *The Yankee Whaler.* Boston: Houghton Mifflin Company, 1926.

514

Bahn, Paul and John Flenley. *Easter Island, Earth Island*. London: Thames and Hudson, 1992.

Baker, William A. *Sloops & Shallops*. Barre, MA: Barre Publishing Co., 1966.

———. *The Lore of Sail*. New York: Facts on File, 1983.

Banks, Charles Edward. *The History of Martha's Vineyard*. Edgartown, MA: Dukes County Historical Society, 1966.

Basseches, Joshua, and Stuart Frank. "Edward Burdett, 1805–1833. America's First Scrimshaw Artist." *Kendall Whaling Museum Monograph Series No. 5*. Sharon, MA: Kendall Whaling Museum, 1991.

Beddie, M. K. *Bibliography of Captain James Cook*. Sydney: Mitchell Library, 1970.

Beechert, Edward D. *Honolulu — Crossroads of the Pacific*. Columbia, SC: University of South Carolina Press, 1991.

Bercaw, Mary K. *Melville's Sources*. Evanston, IL: Northwestern University Press, 1987.

Billingsley, Edward Baxter. *In Defense of Neutral Rights*. Charlotte, NC: University of North Carolina Press, 1967.

Blackburn, Graham. *The Overlook Illus-*

trated Dictionary of Nautical Terms. Woodstock, NY: The Overlook Press, 1981.

Bowen, Abel. *The Naval Monument.* Boston: Bowen, 1816.

Briggs, L. Vernon. *History of Shipbuilding on the North River.* Boston: Coburn Brothers, 1889.

Browne, J. Ross. *Etchings of a Whaling Cruise with Note of a Sojourn on the Island of Zanzibar.* New York: Harper & Brothers, 1846.

Busch, Britton Cooper. *Whaling Will Never Do For Me.* Lexington, KY: University Press of Kentucky, 1994.

Chapelle, Howard I. *The History of the American Sailing Navy.* New York: W. W. Norton, 1949.

Church, Albert Cook. *Whale Ships and Whaling.* New York: W. W. Norton, 1938.

Cleveland, Richard J. *Voyages and Commercial Enterprises of the Sons of New England.* New York: Leavitt & Allen, 1855.

Coletta, Paolo E., ed. *American Secretaries of the Navy.* Annapolis, MD: Naval Institute Press, 1980.

Comstock, William. *Voyage to the Pacific Descriptive of the Customs, Usages and Sufferings on Board of Nantucket Whale-*

Ships. Boston: Oliver L. Perkins, 1838.

————. *The Life of Samuel Comstock, the Terrible Whaleman.* Boston: James Fisher, 1840.

Creighton, Margaret. *Rites and Passages.* Cambridge, U.K.: Cambridge University Press, 1995.

Crevecoeur, Michel-Guillaume Jean de. *Letters from an American Farmer.* London: Westvaco, 1982.

Crosby, Everett U. *Nantucket in Print.* Gloucester, MA: Ten Pound Island Book Company, 1997.

————. *Ninety-Five Per Cent Perfect.* Nantucket, MA: Tetaukimmo Press, 1953.

Dana, R. H. Jr. *The Seaman's Friend.* Boston: Charles C. Little and James Brown and Benjamin Loring and Company, 1841.

Davidson, Louis B. and Eddie Doherty. *Captain Marooner.* New York: Thomas Y. Crowell Company, 1952.

Davis, Captain Charles H. *Life of Charles Henry Davis, Rear Admiral.* New York: Harper Brothers, 1879.

Davis, Charles G. *The Ship Model Builder's Assistant.* New York: Edward W. Sweetman Company, 1970.

Davis, William A. *Nimrod of the Sea.* North

Quincy, MA: Christopher Publishing House, 1972.

de Braganza, Ronald Louis Silveira. *The Hill Collection of Pacific Voyages.* San Diego: University of California, 1974.

Deane, Samuel. *History of Scituate, Massachusetts, From Its First Settlement to 1831.* Boston: Loring, 1831.

Dillon, Peter. *Narrative and Successful Result of a Voyage in the South Seas, to Ascertain the Actual Fate of Lapérouse's Expedition.* London: Hurst, 1829.

Dodge, Ernest S. *New England and the South Seas.* Cambridge, MA: Harvard University Press, 1965.

Douglas-Lithgow, R. A. *Nantucket, A History.* New York: G. P. Putnam's Sons, 1914.

Ely, Ben-Ezra Stiles. *"Thar She Blows": A Narrative of a Whaling Voyage.* Middletown, CT: Wesleyan University Press, 1971.

Emery, Vice Admiral George W. *Historical Manuscripts in the Navy Department Library.* Washington, D.C.: Naval Historical Center, 1994.

Fairburn, William Armstrong. *Merchant Sail.* 6 vols. Gloucester, MA: Ten Pound Island Book Company, 1992.

Farnham, Joseph E. C. *Brief Historical*

Data and Memories of My Boyhood Days in Nantucket. Providence, RI: Snow and Farnham Company, 1923.

Fitzpatrick, Gary L. *The Early Mapping of Hawai'i, Volume I.* Honolulu: Editions Limited, 1986.

Frank, Stuart M. *Dictionary of Scrimshaw Artists.* Mystic, CT: Mystic Seaport Museum, 1991.

————. *More Scrimshaw Artists.* Mystic, CT: Mystic Seaport Museum, 1998.

Goode, George Brown. *The Fisheries and Fishery Industries of the United States. Section IV. The Fishermen of the U.S.* Washington, D.C.: Government Printing Office, 1887.

————. *The Fisheries and Fishery Industries of the United States. Section V. The Whale Fishery.* Washington, D.C.: Government Printing Office, 1887.

Hall, Elton W. *Whalers, Wharves, and Waterways.* New Bedford, MA: Whaling Museum, 1973.

Hall, Henry. *Report on the Ship-Building Industry of the United States.* Washington, D.C.: Government Printing Office, 1884.

Hammatt, Charles H. *Ships, Furs and Sandalwood.* Honolulu: University of Hawai'i Press, 1999.

Harland, John. *Seamanship in the Age of Sail*. Annapolis, MD: Naval Institute, 1984.

Hegarty, Reginald B. *Birth of a Whaleship*. New Bedford, MA: New Bedford Free Public Library, 1964.

Hezel, Francis X. *The First Taint of Civilization*. Honolulu: University of Hawaii Press, 1983.

Hohman, Elmo Paul. *The American Whaleman*. New York: Longmans, Green and Co., 1928.

————. *History of American Merchant Seamen*. Hamden, CT: Shoestring Press, 1956.

Hoyt, Edwin P. *The Mutiny on the* Globe. New York: Random House, 1975.

Huntington, Gale. *Songs the Whalemen Sang*. Barre, MA: Barre Publishers, 1964.

Huntress, Keith. *A Checklist of Narratives of Shipwrecks and Disasters*. Ames, IA: Iowa State University Press, 1979.

Jarves, James Jackson. *History of the Hawaiian Islands Embracing Antiquities, Legends, Discovery by Europeans, Civil, Religious, and Political History*. Honolulu: H. M. Whitney, 1872.

Jenkins, J. T. *A History of the Whale Fisheries*. Port Washington, NY: Kennikat

Press, 1971.

Judd, Walter. *Hawai'i Joins the World.* Honolulu: Mutual Publishing, 1999.

Kemp, Peter, ed. *The Oxford Companion to Ships and the Sea.* London: Oxford University Press, 1976.

King, Dean. *A Sea of Words.* New York: Henry Holt and Company, 1995.

King, Pauline N., ed. *Journals of Stephen Reynolds. Volume I: 1823–1829.* Honolulu: Ku Pa'a Incorporated, 1989.

Kotzebue, Otto von. *A Voyage of Discovery, into the South Sea and Beering's Straits.* . . . 3 vols. London: Longman, Hurst, Rees, Orme, and Brown, 1826.

Kugler, Richard C. "The Penetration of the Pacific by American Whalemen in the 19th Century." *National Maritime Museum Maritime Monographs and Reports, No. 2.* London: National Maritime Museum, 1971.

Kuykendall, Ralph S. *The Hawaiian Kingdom. 1788–1854.* Honolulu: University of Hawaii Press, 1947.

Lay, William and Cyrus Hussey. *A Narrative of the Mutiny on Board the Whaleship* Globe. New York: Corinth Books, 1963. (Reprint of 1828 original.)

Leach, Robert J. and Peter Gow. *Quaker Nantucket.* Nantucket, MA: Mill Hill Press, 1997.

Leavitt, John F. *The Charles W. Morgan.* Mystic, CT: Mystic Seaport, 1973.

Levy, Neil M. *Micronesia Handbook.* Emeryville, CA: Avalon Travel Publishing, 2000.

Lofstrom, William L. *Paita: Outpost of Empire.* Mystic, CT: Mystic Seaport Museum, 1996.

Long, David F. *"Mad Jack."* Westport, VT: Greenwood Press, 1993.

Lund, Judith Navas. *Whaling Masters and Whaling Voyages Sailing from American Ports.* Gloucester, MA: Ten Pound Island Book Company, 2001.

Lytle, Thomas G. *Harpoons and Other Whalecraft.* New Bedford, MA: Whaling Museum, 1984.

Macy, Obed. *The History of Nantucket.* Mansfield, MA: Macy & Pratt, 1880.

Macy, William F. *The Nantucket Scrap Basket.* Boston: Houghton Mifflin Company, 1930.

Malone, Dumas, ed. *Dictionary of American Biography.* New York: Charles Scribner's Sons, 1936.

Maloney, Linda M. *The Captain from Connecticut. The Life and Times of Isaac Hull.*

Boston: Northeastern University Press, 1986.

Maury, M. F. *Explanations and Sailing Directions to Accompany the Wind and Current Charts.* . . . Philadelphia: E. C. and J. Biddle, 1855.

McKay, Richard C. *South Street.* New York: G. P. Putnam's Sons, 1934.

McLaren, I. F. *Laperouse in the Pacific.* Parkville, Australia: University of Melbourne Library, 1993.

Meade, Cdr. R. W. "Admiral Hiram Paulding." *Harper's New Monthly Magazine.* LVIII (1879): 358–364.

Melville, Herman. *Moby-Dick.* New York: Library of America, 1983.

———. *Typee.* New York: Library of America, 1982.

Moebs, Thomas Truxtun. *America's Naval Heritage.* Washington, D.C.: Naval Historical Center, 2000.

Mooney, Robert F. and Andre R. Sigourney. *The Nantucket Way.* Garden City, NY: Doubleday and Company, Inc., 1980.

Morison, Samuel Eliot. "Historical Notes on the Gilbert and Marshall Islands." *The American Neptune IV* (1944): 87–118.

———. *The Maritime History of Massachu-*

setts. Boston: Houghton Mifflin Company, 1921.

———. *The Oxford History of the American People*. New York: Oxford University Press, 1965.

Morrell, Benjamin. *A Narrative of Four Voyages, to the South Sea and South Pacific Ocean Chinese Sea, Ethiopic and Southern Atlantic Ocean, Indian and Antarctic Ocean*. New York: J. & J. Harper, 1832.

New England Historical and Genealogical Society. *Vital Records of Nantucket Massachusetts to the Year 1850*. Salem, MA: Essex Institute, 1925.

Olmstead, Francis Allyn. *Incidents of a Whaling Voyage*. Rutland, VT: Charles E. Tuttle Company, 1969.

Paulding, Hiram. *Journal of a Cruise of the U.S. Schooner Dolphin*. . . . New York: G. & C. & H. Carvill, 1831.

Paullin, Charles Oscar. *Diplomatic Negotiations of American Naval Officers. 1778–1883*. Baltimore: Johns Hopkins, 1912.

Philbrick, Nathaniel. *Away Off Shore*. Nantucket, MA: Mill Hill Press, 1994.

———. *In the Heart of the Sea*. New York: Viking, 2000.

Porter, David. *Journal of a Cruise Made to the Pacific Ocean*. 2 vols. New York:

Wiley and Halsted, 1822.

Robinson, J. H. *Guide to Nantucket.* Nantucket, MA: J. H. Robinson, 1948.

Ronnberg, Erik A. R. Jr. *To Build a Whaleboat.* New Bedford, MA: Old Dartmouth Historical Society Whaling Museum, 1985.

Sacks, Oliver. *The Island of the Colorblind.* New York: Vintage Books, 1998.

Skallerup, Harry R. *Books Afloat & Ashore.* Hamden, CT: Archon Books, 1974.

Snow, Caleb H. *A History of Boston.* Boston: Abel Bowen, 1825.

Snow, Edward Rowe. *Piracy, Mutiny and Murder.* New York: Dodd, Mead and Company, 1959.

Snow, Philip and S. Waine. *The People from the Horizon.* Oxford, U.K.: Phaidon Press, 1979.

Stackpole, Edouard A. *Mutiny at Midnight.* New York: William Morrow and Company, 1939.

———. *The Mutiny on the Whaleship Globe.* (No place of publication given): Author, 1981.

———. *The Sea-Hunters.* Philadelphia: J. P. Lippincott Company, 1953.

———. *Whales & Destiny.* (No place of publication given): University of Massachusetts Press, 1972.

Starbuck, Alexander. *History of the American Whale Fishery.* Washington, D.C.: Government Printing Office, 1878.

———. *The History of Nantucket.* Rutland, VT: Charles E. Tuttle Company, 1969.

Stevens, William Oliver. *Nantucket.* New York: Dodd, Mead and Company, 1941.

Stewart, C. S. *A Visit to the South Seas in the United States Ship* Vincennes. 2 vols. London: Henry Colburn and Richard Bentley, 1832.

Story, Dana A. *The Building of a Wooden Ship.* Barre, MA: Barre Publishers, 1971.

———. *Frame Up!* Gloucester, MA: Ten Pound Island Book Company, 1986.

———. *The Shipbuilders of Essex.* Gloucester, MA: Ten Pound Island Book Company, 1995.

Tower, Walter S. *A History of the American Whale Fishery.* Philadelphia: For the University, 1907.

"Trustum" and His Grandchildren. Nantucket: Author, 1881.

Turner, Harry B. *Nantucket Argument Settlers.* Nantucket, MA: Author, 1946.

Ward, R. Gerard, ed. *American Activities in the Central Pacific. 1790–1870.* 8 vols. Ridgewood, NJ: Gregg Press, 1967.

Whipple, A. B. C. *Yankee Whalers in the South Seas.* Garden City, NY: Doubleday and Company, Inc., 1954.

Wilkes, Charles. *Narrative of the United States Exploring Expedition.* 5 vols. Philadelphia: Sherman, 1849.

Williams, Winston. *Nantucket Then and Now.* New York: Dodd, Mead and Company, 1977.

Manuscript collections at:
Alele Library, Majuro, Marshall Islands
Cincinnati Historical Society
Martha's Vineyard Historical Society
Nantucket Historical Association
National Archives
Navy Department Library

Acknowledgments

My primary readers and coconspirators in this project were Anthony Weller, Llewellyn Howland, Stuart Frank, and John Brown. Erik Ronnberg, Dana Story, Mary K. Bercaw Edwards, Glenn Grasso, George Emery, and Dr. James McGee scoured various sections of the text for my frequent and lubberly errors, and Dr. McGee very generously worked with me toward some understanding of Samuel Comstock's mental state. Anne Marie Crotty, Joe Burns, and Orville Haberman were kind enough to point out the parts of the narrative that confused them or put them to sleep. Agent Neeti Madan and editor Deborah Baker were patient friends and gentle teachers who kept me from wandering too far in my journey from one end of the narrative to the other.

There was no telling this story without the help of the people who guided me through its sources. Tom Weaver and J. C.

and Janice Ramsey initiated me to the mysteries of Vevay, Indiana, and the location of Ted Langstroth's book hoard. Fellow book dealers Owen Kubik and John Mullins saw Strong's manuscript for what it was, and brought it to my attention. Stuart Frank and Michael Dyer of the Kendall Whaling Museum made the manuscript, and all the resources of that excellent institution, available to me. Betsy Lowenstein and the staff of the Nantucket Historical Association similarly guided me through the materials in their care, including Edouard Stackpole's own collection of manuscripts and documents relating to the *Globe* mutiny. Matthew Stackpole and the staff at the Martha's Vineyard Historical Society — especially Jill Bouck, Arthur Railton, and Kay Mayhew — helped me find some wonderful material. Rennie Stackpole of the Penobscot Marine Museum provided a beautiful copy of the *Globe*'s crew list, an important document for which I had searched in vain. Will LeMoye and the top-notch staff at the Peabody Essex Museum opened their doors to me, as did Jean Hort and Glenn Helm at the Navy Department Library, Charles Sparrell at the Scituate Historical Society, and David

Brown at the Essex Shipbuilding Museum. My thanks also to Judy Lund for combing the files of the New Bedford Whaling Museum in search of new information about the *Globe,* and to colleague Ed Lefkowicz who made me aware of the rare Wisconsin edition of Lay's narrative.

Suzanne Murphy of the Marshall Islands Visitors Authority helped us arrange our journey to Mili Atoll, and Mr. and Mrs. Chutaro helped us understand it. On Mili, Minister Tadashi Lometo very generously provided us with a place to stay, and the wonderful people of the islands made us feel comfortable and welcome. Thank you Jinwa, Tamar, Jane, Moses, Gary, Junior, and Norio. When you visit me, I'll show you my island.

About the Illustrators

ERIK RONNBERG is a model-maker, marine draftsman, and maritime historian. He is former assistant curator of the New Bedford Whaling Museum and former editor of the *Nautical Research Journal*. He now lives in Beverly, Massachusetts, where he is a freelance model-maker represented exclusively by the American Marine Model Gallery in Salem, Massachusetts.

GARY TONKIN is a scrimshaw artist and book illustrator living in the western Australian whaling port of Albany. He has had firsthand experience in the modern whaling industry and is active in the preservation of Australia's whaling heritage. He is a member of the Australian Society of Marine Artists, and from 1999–2001 was the artist in residence at the Kendall Whaling Museum in Sharon, Massachusetts.